TABLE OF CONTENTS

SUMMARY OF DESIGN DATA

	Weight lb	Ventilation, cfm/hd Cold weather rate	Mild weather rate	Hot weather rate	Winter room temperature F	Supplemental heat, Btu/hr/hd Slotted floors	Bedded or scraped floors	Manure, ft³/hd Liquids + solids + 15% extra
Sow and litter	400	20	80	500	80	4000	—	0.66
					70	3000	—	
					60	—	3500	
Prenursery pig	12-30	2	10	25	85	350	—	0.03
Nursery pig	30-75	3	15	35	75	350	—	0.07
					65	—	450	
Growing pig	75-150	7	24	75	60	600	—	0.14
Finishing pig	150-220	10	35	120	60	600	—	0.24
Gestating sow	325	12	40	150*	60	1000	—	0.20
Boar	400	14	50	300	60	1000	—	0.25

*300 cfm for gestating sows in a breeding facility.

Slot Widths

For slatted floors. Wire mesh, metal, or plastic slats preferred in farrowing and prenursery.

	Slot widths in.	Concrete slat widths, in.
Sow and litter	3/8	4
Prenursery pig	3/8	Not recommended
Nursery pig	1	4
Growing-finishing pig	1	6-8
Gestating sows or boars:		
Pens	1	6-8
Stalls	1	4

Water Requirements

Animal type	Gal/hd/day
Sow and litter	8
Nursery pig	1
Growing pig	3
Finishing pig	4
Gestating sow	6
Boar	8

Space Requirements

Enclosed housing:

Pigs	Weight lb	Area ft²
Prenursery[a]	12-30	2-2½
Pig-nursery[b]	30-75	3-4
Growing[b]	75-150	6
Finishing[b]	150-220	8

[a]Avoid concrete slats, slats over 2″ wide, and partly slotted floors for prenursery pigs.
[b]For slotted, flushed, or scraped floor.

Shed With Lot:

More lot area is often provided to facilitate manure drying.

	Weight lb	Inside ft²/hd	Outside ft²/hd
Nursery pig	30-75	3-4	6-8
Growing/finishing pig	75-220	5-6	12-15
Gestating sow	325	8	14
Boar	400	40	40
Sow in breeding	325	16	28

Breeding swine	Weight lb	Solid floor ft²	Totally or partly slotted floor[a] ft²	Animals per pen	Stall size
Breeding					
Gilts	250-300	40	24	up to 6	
Sows	300-500	48	30	up to 6	
Boars	300-500	60	40	1	2′- 4″x7′
Gestating					
Gilts	250-300	20	14	6-12	1′-10″x6′
Sows	300-500	24	16	6-12	2′- 0″x7′

[a]or flushed open gutter. Open gutter not recommended in breeding because of slick floors.

Feeder Space:

Sows: 1′/self-feed sow, 2′/group-fed sow.
Pig (12-30 lb): 2 pigs/feeder space
Pig (30-50 lb): 3 pigs/feeder space
Pig (50-75 lb): 4 pigs/feeder space
Pig (75-220 lb): 4-5 pigs/feeder space

Space Requirements

Pasture Space:
Depends on rainfall and soil fertility.

10 gestating sows/acre
7 sows with litters/acre
50 to 100 growing-finishing pigs/acre

Shade Space:

15-20 ft²/sow
20-30 ft²/sow and litter
4 ft²/pig to 100 lb
6 ft²/pig over 100 lb

Waterer Space:

Minimum of 2 waterers per pen.
Pig (12-75 lb): 10 pigs/waterer
Pig (75-220 lb): 15 pigs/waterer

Floor and Lot Slopes

Slotted floors: usually flat
Solid floors:
Farrowing:
¼"-½"/ft without bedding
¼"/ft with bedding

Pigs:
½"/ft without bedding
¼"/ft with bedding

Paved lots: ½"/ft
Paved feeding floors:
Indoors: ¼"/ft
Outdoors: ½"/ft

Building alleys:
½"/ft crown or side slope
⅛"/ft to drains

CONVERSIONS

AREAS & VOLUMES

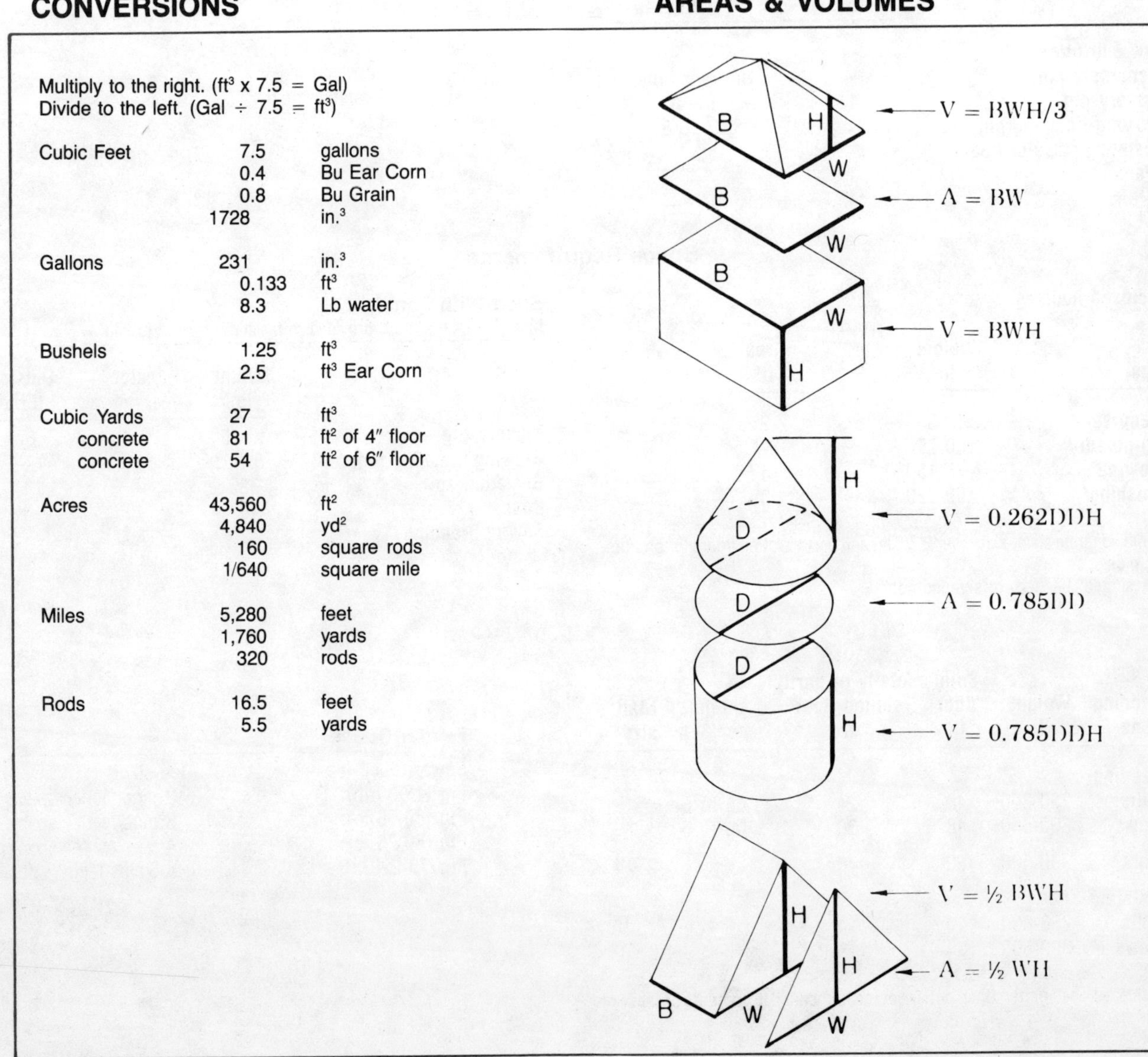

Multiply to the right. (ft³ x 7.5 = Gal)
Divide to the left. (Gal ÷ 7.5 = ft³)

Cubic Feet	7.5	gallons
	0.4	Bu Ear Corn
	0.8	Bu Grain
	1728	in.³
Gallons	231	in.³
	0.133	ft³
	8.3	Lb water
Bushels	1.25	ft³
	2.5	ft³ Ear Corn
Cubic Yards	27	ft³
concrete	81	ft² of 4" floor
concrete	54	ft² of 6" floor
Acres	43,560	ft²
	4,840	yd²
	160	square rods
	1/640	square mile
Miles	5,280	feet
	1,760	yards
	320	rods
Rods	16.5	feet
	5.5	yards

This handbook summarizes current agricultural engineering recommendations for swine producers. It deals with the design and operation of buildings and the equipment necessary for an efficient swine operation.

Many of the example buildings in this handbook have a "mwps"-number (e.g. mwps-72682), which indicates that blueprints are available. Contact your Extension Agricultural Engineering department (see inside front cover) or the Midwest Plan Service.

SITE SELECTION FOR SWINE FACILITIES

Many factors determine the "best" site for swine facilities. While most are a matter of common sense, many operations are poorly located because some important factor was overlooked. This section is primarily for producers planning new facilities.

Space

It is wise to assume your enterprise will double in size and plan accordingly. You need space for buildings, clearance between buildings (at least 35′ for most buildings and 50′ for naturally ventilated buildings), and to enlarge buildings. You need lanes for vehicle access and room for parking. Allow for a feed center and adequate separation from family housing and other farm activities.

Water

A year-round supply of potable water is essential for watering animals and sanitation. Water is also needed for fire protection and lagoon dilution.

A modern operation is a large water user—a 1000-head/yr farrow-to-finish operation uses 2500 gal/day. Where ground water supplies are not adequate, use surface sources such as farm ponds or community water systems. The extra expense of surface sources may justify selecting another site. See section on utilities.

Drainage

Construct outside lots so rainfall and snowmelt run off without backing up onto the lot. Provide a 4% slope on outside lots. Divert upslope runoff around the lots to reduce the amount of contaminated runoff to be handled. A minimum of 1% slope from lots or buildings to lagoons and holding ponds allows gravity transport of runoff or liquid manure. High water tables affect construction of buildings, lagoons, and manure storages.

Residential Housing

Locate swine facilities for easy access by farm labor, but isolate their dust, noise, and odor from the family living center. In small to medium operations, where swine managers spend much of their working day elsewhere, place the facilities convenient to the farm service area. As a rule, locate high-labor facilities, such as farrowing and nursery buildings, at least 300′ away from the family living area. Put the large animal facilities, such as finishing and gestation, at least 500′ away. Locate larger operations with full-time personnel even farther away from the living area because of the greater dust and odor problems and because labor movement back and forth to the rest of farmstead is not as important.

Production volume and direction of prevailing summer breezes affect location of swine facilities with respect to neighbors. Odors from swine units are often detectable ½ mile downwind. Odors from large swine units are noticeable up to 3 miles away.

Some practices affect the frequency of odor complaints. Open lot systems and irrigation or broadcast spreading of liquid manure generate more odor than environmentally controlled buildings and soil-incorporated manure application. Overloaded or improperly managed lagoons also cause odor nuisances.

Some areas have legally established separation distances. Regulations may restrict locating swine operations near:

- Neighboring residences.
- Public buildings and recreational areas.
- Built-up residential areas.

What your neighbors **see** affects their attitude toward swine operations. A neat, well-kept operation usually receives fewer complaints than those in disrepair. Use visual screens, such as trees or buildings, to hide lagoons and other "unsightly" areas that remind people of odor.

Existing Facilities

Don't let an existing swine building be the major factor in locating a large operation of considerable investment. An existing building at a remote location can often serve as a gestation building, an overflow finishing building, or an isolation unit for purchased animals. See section on existing buildings.

An access lane can usually be economically replaced. The cost of a new well or extensive plumbing, for example, can often be justified by the advantages of the best site. Generally, it is best to locate a new swine operation relative to more permanent items, such as public roads and residential housing, then consider existing wells, access lanes, and buildings.

Manure Management

Select a building site with sufficient land for spreading manure. The minimum acreage required by many state pollution agencies is based on satisfying the nitrogen requirements of the growing crop; however, a buildup of phosphorus can limit crop yields in some cases. Manure handling practices which result in a loss of nitrogen, such as lagooning and aeration, decrease the amount of land required. You may be able to acquire manure spreading rights from adjacent land owners.

The land must be suitable for spreading. Avoid steep slopes where manure runoff can cause water pollution, and avoid land adjacent to neighboring residences.

Snow and Wind Control

Arrange buildings and lots to minimize drifting problems. Provide at least 35′ between buildings to allow space for snow catch and stockpiling. Place access lanes out of depressions and other snow drop areas. Locate naturally ventilated buildings and open lots so they receive undisturbed summer breezes.

Feed Supply

Feed supply location is a consideration, but not an overriding concern, in choosing a site. Transport wagons can satisfactorily link a grain-feed center to a swine operation. Many producers purchase commercial feed and store it in bulk bins at the swine operation. Large swine units may have a satellite feed mixing center in addition to the main grain center. See section on grain-feed centers.

Security

Consider theft, vandalism, and fire safety. Limit farm visitor access to control disease and to reduce interference with farm work. If located on the same farmstead as the manager's residence, run the swine center access lane near the home. If a second access is used for feed, manure, and animal transport vehicles, provide an alarm system to guard against unauthorized traffic.

Facilities remote from the manager's residence pose the most problems. Provide only one access road—unauthorized persons are less apt to visit if there is no escape route should the manager return. If possible, make access roads at remote sites visible from a public road or neighboring residence.

Plans, Specifications, and Contracts

Detailed plans, specifications, and contracts help provide the needed communication and understanding between owner and builder so that you get what you want.

Complete **plans** show all the necessary dimensions and details for construction. **Specifications** support the plans; they describe the materials to be used, including size and quality, and often outline procedures for construction and quality of workmanship. The **contract** is an agreement between the builder and the owner; it includes price of construction, schedule of payments, guarantees, responsibilities, and starting and completion dates.

You have several options for preparing this material:

Be your own contractor. Draw a final plan, making sure dimensions are correct and construction details and materials are determined. Have plans checked by the appropriate regulatory agency when required. Determine total costs before beginning construction.

Hire a consulting engineer to prepare final working drawings, prepare specifications, and supervise bid letting to contractors.

Use a design and construction firm. Some firms make working drawings and specifications and have standard contract forms.

Consulting Engineers

Services commonly offered by consulting engineers include:

- Direct personal service (technical advice, etc.).
- Preliminary investigations, feasibility studies, and economic comparison of alternatives.
- Planning studies.
- Design.
- Cost estimates.
- Engineering appraisals.
- Bid letting.
- Construction supervision and inspection.

Consulting engineers usually do a project in three phases: preliminary planning, engineering design, and construction monitoring. They may be retained to help with one or more of these phases. To select a consulting engineer, consider:

- Registration: To protect the public welfare, states certify and license engineers of proven competence. Practicing consulting engineers must be registered professional engineers in their state of residence, and qualified to obtain registration in other states where their services are required.
- Technical qualifications.
- Reputation with previous clients.
- Experience on similar projects.
- Availability for the project.

SCHEDULING

Facility scheduling is one of the most important and often the most difficult task in managing an intensive production system. Production system capacity is usually based on desired farrowing frequency, sow herd size, farrowing building size, or number of pigs finished per year. Once you have a value for two or more of these factors, determine **general** facility sizes from Table 1. Make a separate **detailed** scheduling analysis of your particular system before starting construction.

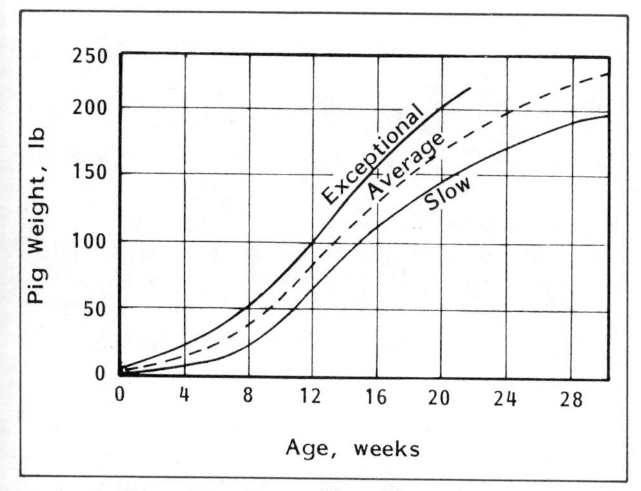

Fig 1. Typical growth of swine.

Table 1 shows the average number of sows, boars, and pigs in various production schedules, based on a 10-stall farrowing building (40 stalls for a weekly schedule). For a larger or smaller farrowing building, calculate a multiplier by dividing the desired number of farrowing stalls by the table value. Apply the multiplier to the table entries in the same row as the desired farrowing frequency. To base the capacity on number of pigs produced per year, divide desired number of pigs per year by the table value to get the multiplier.

Example A

A producer wants to farrow 7 times a year in a 24-stall farrowing building. Determine the size of the other facilities and the number of finishing hogs produced per year.

Row 2 in Table 1 is for 7 farrowings per year. Calculate the multiplier by dividing the desired number of farrowing stalls by the table value: 24 ÷ 10 = 2.4. Multiply the other values in row 2 by 2.4:

- sows in entire herd = 2.4 x 39 = 93.6 (round up to 94).
- sows in gestation unit = 2.4 x 24 = 57.6 (58).
- boars in breeding unit = 2.4 x 5 = 12.
- pigs in nursery unit = 2.4 x 96 = 230.4 (231).
- pigs in growing unit = 2.4 x 80 = 192.
- pigs in finishing unit = 2.4 x 80 = 192.
- pigs marketed per year = 2.4 x 595 = 1428.

Table 1. Swine scheduling table.
Animal capacity per 10 sows farrowed per period.

						Use "Multiplier" on these values:								
						Sow capacity in:				Pig capacity in:				
Row	Farrowing frequency A	Farrowing interval, weeks B	Number of sow groups C	Farrowing stalls D	Total sows E	gestation	gilt pool	breeding	Boars	prenursery	nursery	growing	finishing	Pigs per year
						- - - - - - -F- - - - - -				- - - - - - - - - -G- - - - - - - - -				
1	6x	8	3	10	39	24	12	15	4	—	96	—	160	520
2	7x(1st)	7	3	10	39	24	12	15	5	—	96	80	80	595
3	8x	6	4	10	51	36	12	15	4	—	96	80	160	695
4	10x(1st)	5	4	10	51	36	12	15	5	—	96	160	160	830
5	12x(1st)	4	5	10	63	48	12	15	6	96	80	160	160	1040
6	Weekly(1st)	4	20	40	243	168	12	45	7	340	320	640	680	4160

A. 6x = farrowing about 6 times per year.
(1st) = sows bred during first heat after weaning.

B. Time per sow group for farrowing and cleaning.

C. Weekly farrowing requires 5 sow groups for each of 4 rooms = 20 groups.

D. Group size = number of farrowing stalls. Weekly farrowing requires at least 4 rooms of 10 stalls each = 40 stalls. Five rooms (50 stalls) are preferred to allow one week for cleaning between farrowing.

E. Sows = sows + gilts. Gilt pool not included.
Total sows = (group size x 1.2) x (number of sow groups + ¼). The 1.2 allows for 80% conception. The ¼ allows for open sows returned from gestation.

F. Building capacity needed—not necessarily number of sows at all times. In most cases, space for one group is needed only when cleaning farrowing room and is not used after the group is moved to farrowing. It may be possible to provide short term housing for this group elsewhere on the farm. Gilt pool = 6 x typical number of sows to be replaced in each sow group (20% replacement assumed). Breeding area allows extra capacity for low conception rates during warm weather.

G. Assumes 8 pigs marketed per litter. Pigs marketed at the average age of 26 weeks. Capacity for first group moved to prenursery or nursery is increased by 20% to account for overfarrowing and slow growers.

Example B

A producer wants a new weekly farrowing complex. He needs to produce about 5000 pigs per year to utilize his labor and grain.

Row 6 is for weekly farrowing. Determine the multiplier by dividing the desired number of pigs per year by the table value: 5000 ÷ 4160 = 1.2. Multiply the other values in row 6 by 1.2:

- farrowing stalls = 1.2 x 40 = 48 (12 stalls per room).
- sows in entire herd = 1.2 x 243 = 291.6 (round up to 292).
- sows in gestation unit = 1.2 x 168 = 201.6 (202).
- boars in breeding unit = 1.2 x 7 = 8.4 (9)
- pigs in prenursery unit = 1.2 x 340 = 408.
- pigs in nursery unit = 1.2 x 320 = 384.
- pigs in growing unit = 1.2 x 640 = 768.
- pigs in finishing unit = 1.2 x 680 = 816.
- pigs marketed per year = 1.2 x 4160 = 4992.

Example C

A producer has two 20-sow farrowing rooms. He wants to farrow one sow group every 3 weeks (i.e. each room will be occupied for 6 weeks per farrowing). This is equivalent to two separate 20-sow farrowing systems, so figure the capacity of one room and double it.

Row 3 is for 6 weeks per farrowing. The multiplier is the actual number of sows per room divided by 10: 20 ÷ 10 = 2.0. Multiply the other values in row 3 by 2.0:

- number of sow groups = 4 per room (8 for two rooms).
- total sows = 2.0 x 51 = 102 (204 for two rooms).
- sows in gestation = 2.0 x 36 = 72 (144 for two rooms).
- sows in breeding = 2.0 x 15 = 30 (60 for two rooms).*
- boars needed = 2.0 x 4 = 8 (8 for two rooms).*
- pigs in nursery = 2.0 x 96 = 192 (384 for two rooms).
- pigs in growing = 2.0 x 80 = 160 (320 for two rooms).
- pigs in finishing = 2.0 x 160 = 320 (640 for two rooms).
- pigs produced per year = 2.0 x 695 = 1390 (2780 for two rooms).

* This example assumes second heat breeding (3½ weeks after weaning), so space for two groups of sows is required in breeding building. However, the two sow groups are bred 3 weeks apart, so you need only enough boars for one sow group.

FARROWING FACILITIES

Typical Farrowing Schedules

- **One Litter per Year.** Gilts are farrowed once a year in mild weather, usually on pasture. Building and equipment investment must be small, because the cost is charged to only one group of sows and litters per year.
- **Two Litters per Year.** One group of sows is farrowed twice a year, often on pasture. Farrowings are scheduled to avoid weather extremes and periods of high labor demand. Overhead for buildings and equipment is prorated to two litters per year.
- **Multiple Litters per Year.** Two or more groups of sows farrow twice a year, allowing 4 to 12 farrowings a year. Many producers farrowing 8 or more times per year end up with continuous farrowing because of the difficulty in getting groups of sows to farrow over a short time. Environmentally controlled facilities are common for this schedule, which increases overhead costs but reduces labor per pig. Facility costs are charged to many pigs, lowering the cost per pig.
- **Farrowing Each Week.** Some large producers farrow a set number of sows (usually 8 to 12) each week in separate, small farrowing rooms, Fig 2. Weekly farrowing provides better overall management and full use of specialized labor and facilities, resulting in groups of sows bred weekly, litters born within the same week, and rooms that can be emptied and disinfected between groups. Provide at least 4 farrowing rooms for 3-week weaning—5 rooms for 4-week weaning. An additional room allows time for more convenient cleanup and space for farrowing excess or late sows.

Building Design Factors

Farrowing units vary from a portable A-frame on pasture to an environmentally controlled central building. When selecting a farrowing unit, consider available capital, amount and type of labor, anticipated future size of operation, existing buildings, and management ability. Generally, the more intense farrowing schedules require more capital and management ability, but less labor per pig.

Temperature

Sows and young nursing pigs require different temperatures. A newborn pig needs a dry, draft-free environment at a temperature of 90-95 F the first three days of life. In contrast, a sow is most comfortable at 60-65 F.

Provide these two temperatures by keeping the farrowing building at 65-75 F and providing supplemental heat in the creeps. Creep heaters may be lamps, catalytic gas heaters, electric floor pads, electric heating cables, hot water pipes in the floor, or some combination. See section on ventilation. Provide creep heat for at least the first week after birth. Room temperature can be 60 F if bedding is used. If the building is not environmentally controlled, use draft barriers, bedding, and hovers.

The attendant must recognize signs of stress in the animals and make appropriate adjustments. Pigs pile up or shiver when cold and avoid heaters that are set too warm or too close to the floor.

Sanitation

Properly designed farrowing units are easy to keep clean. Avoid porous or rough surfaces that can harbor bacteria. Rough floors are abrasive to the feet

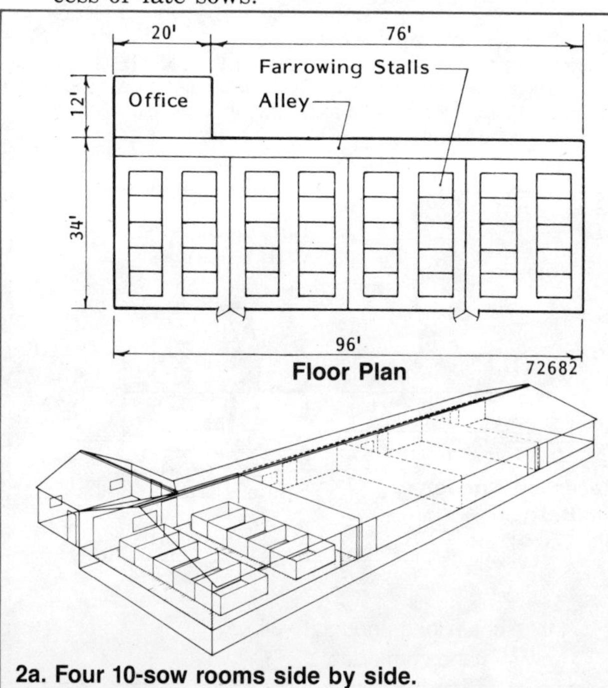

2a. Four 10-sow rooms side by side.
This plan allows good isolation and sanitation management. mwps-72682, option A.

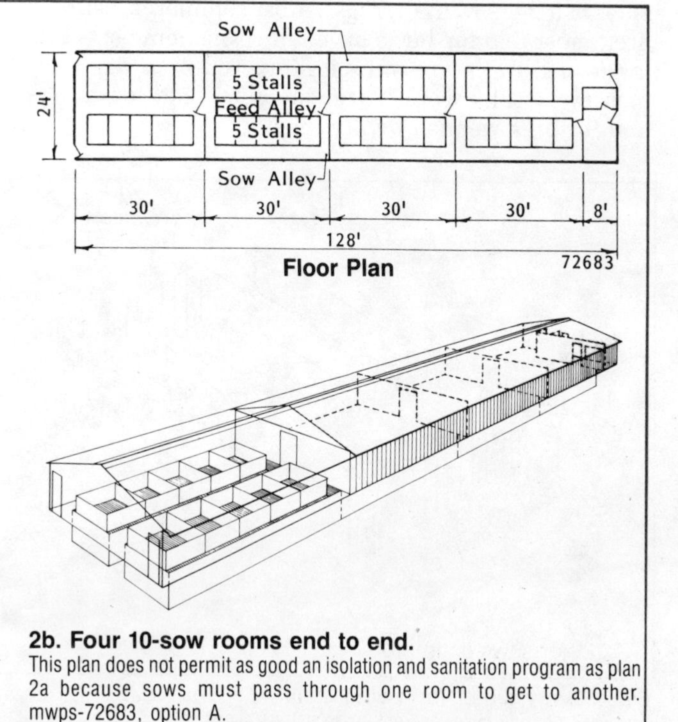

2b. Four 10-sow rooms end to end.
This plan does not permit as good an isolation and sanitation program as plan 2a because sows must pass through one room to get to another. mwps-72683, option A.

Fig 2. Multi-room farrowing buildings.

and knees of nursing pigs and retain moisture and manure. Smooth surfaces drain and dry more rapidly, are easier to clean and disinfect, but can be slippery. Slick floors also result in injured feet and legs. Avoid excessive troweling of concrete floors—power troweling frequently results in slippery floors.

Slope solid floors for proper drainage. Slats made of slick materials (aluminum, stainless steel, plastic) should have ribs down the center or punched slots for better footing. Avoid slats with continuous, parallel ribs which can trap moisture and manure, resulting in unsanitary conditions.

Establish a schedule for breeding and follow it so you can completely empty the building and clean and disinfect between sow groups.

Stalls or Pens

The main differences between farrowing stalls and pens are the amount of exercise for the sow and the degree of protection of the pigs from crushing. With stalls, pigs are protected from the sow. Heated creeps further protect pigs by attracting them away from the sow when not nursing. Stalls require less bedding and labor than do pens. It is more difficult to catch pigs in a stall but the operator is protected from the sow when handling pigs.

Farrowing pens allow more sow movement but require more cleaning labor and floor space. A sow in an open pen must be restrained for any physical treatment. Pens may be preferred for late weaning to give the larger pigs more space. Convert farrowing pens with feeders and waterers to small nursery pens by removing the guard rails.

Stalls are usually 5′ wide by 7′ long (some are 6′ or 6½′ long with the feeder mounted on the outside of the stall). The width includes an 18″ pig area on both sides of a 24″ sow stall, Fig 3. Most commercial stalls are adjustable for large or small sows. Some special farrowing stalls are only 4½′ or even 4′ wide. The narrower stalls work better for gilts, smaller sows, and for pigs weaned by 3 weeks.

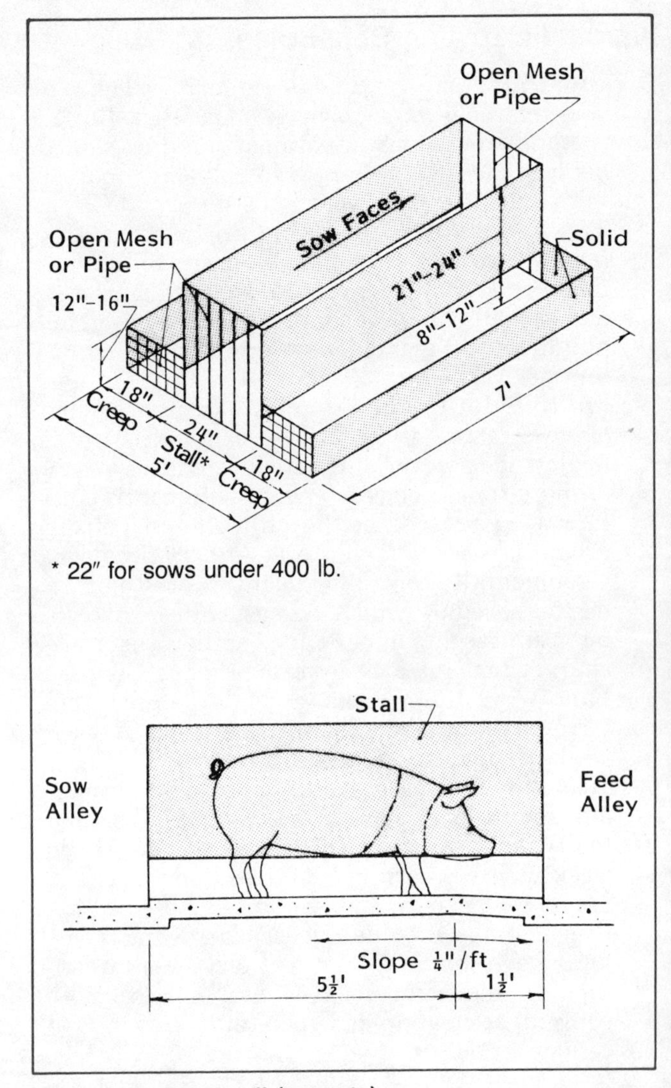

Fig 3. Farrowing stall (or crate).
Farrowing stalls are often 1″ lumber, ¾″ exterior plywood, or 1″ galvanized or aluminum pipe. Solid partitions reduce drafts. Make the front and side creep panels solid. Make the rear creep panels and the front and rear sow stall panels open mesh or pipe.

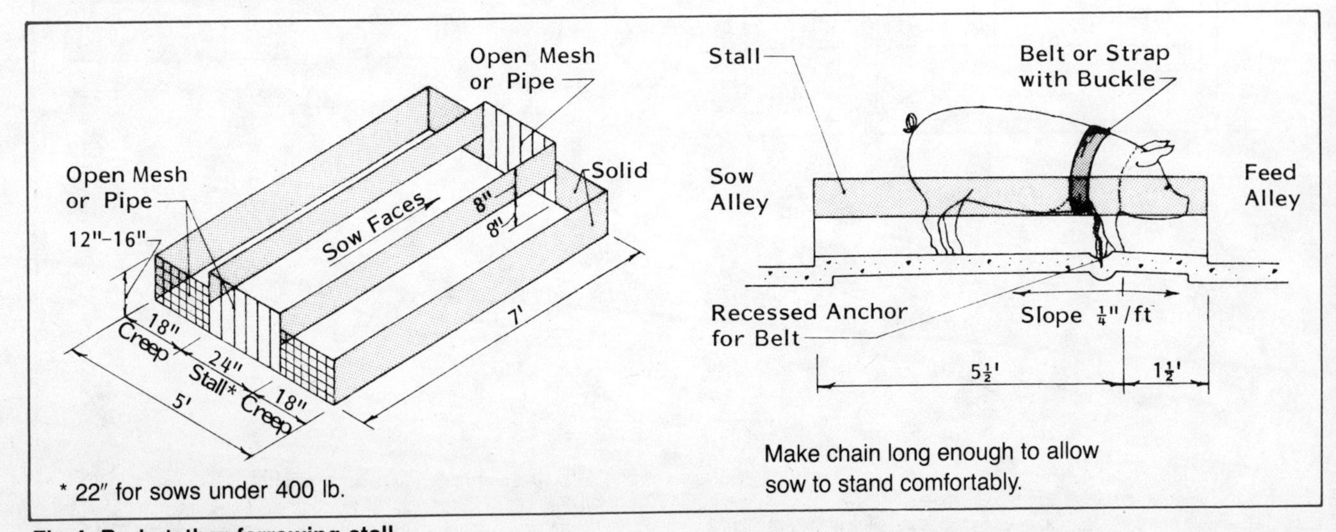

Fig 4. Body tether farrowing stall.
Sow neck or body tethers permit lower stall partitions (12″-18″) for easier observation.

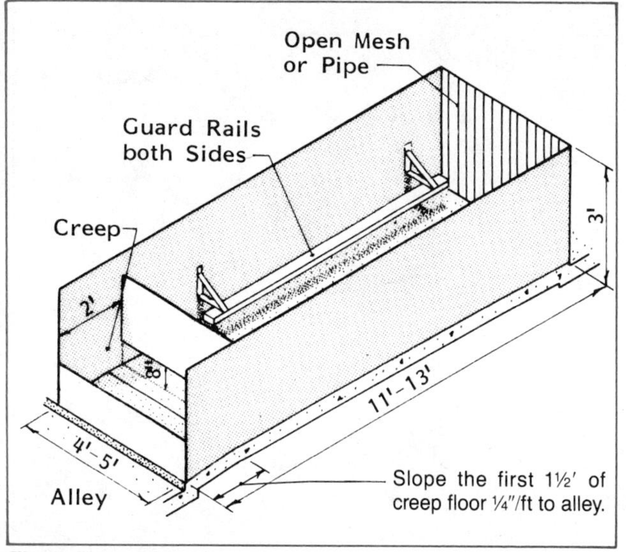

Fig 5. Farrowing pen.
The two-slope floor allows drainage from the creep area directly to the alley instead of across the pen area.

Typical pens are 4′ to 5′ wide by 11′ to 13′ long, Fig 5. The front 18″ to 24″ of the pen is the creep area. Guard rails, about 6″ out from the wall and 8″ off the floor, reduce pig crushing. The rear 4′ to 5′ is often slotted. Cover slats during, and for the first few days after, birth. Locate waterer over slats.

Some producers use farrowing pens as "free stalls." Sows are free to come and go from the pens to a feeding floor for feed and water. Keep pigs in the pens with a low gate behind the sow, Fig 6.

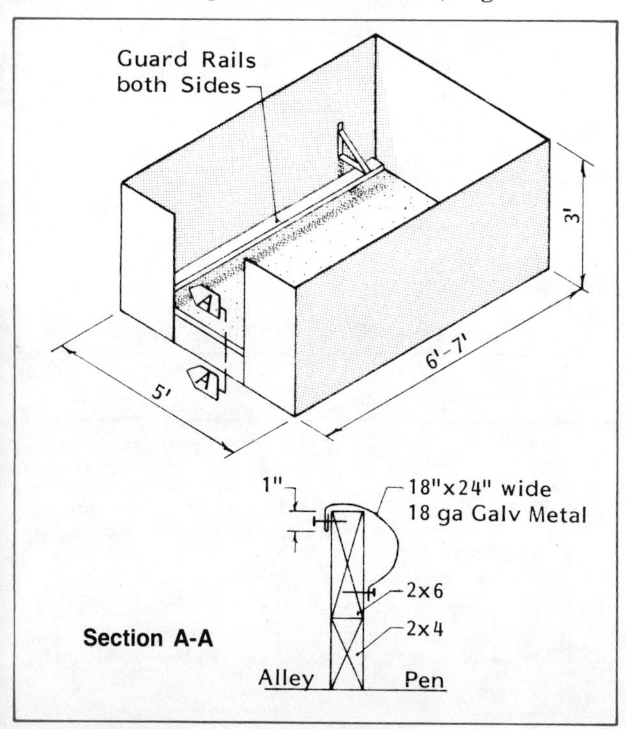

Fig 6. Free stall.
One type of homemade pig stop is shown; commercial units are available.

Floors

Slotted floors

Slotted floors greatly reduce labor to remove manure. They separate the animal from its manure and reduce internal parasite problems. They also result in drier floors.

Buildings with totally slotted floors require more environmental control than buildings with partly slotted or solid floors. Use a solid covering in the creep area of totally slotted floors for about one week after farrowing to increase pig comfort and prevent hoof injuries.

Slat spacing is critical in a farrowing unit. Space slats either a uniform ⅜″ apart **or** 1″ apart. Pig's legs may get caught in spaces **between** ⅜″ and 1″. Cover any spacing over ⅜″ during, and for 3 days after, farrowing. Space concrete and wood slats 1″ apart behind the sow to improve cleaning. A 2″x4″ opening behind the sow is useful for manual manure removal, but keep it covered except during use. Maximum slat width is 4″.

Place slats parallel to the sow to provide better footing for the sow to get up and for the pigs while nursing. Do not install a step from the slotted to solid floor areas of partly slotted floors under sows in stalls. See Fig 7 for several farrowing stall slat arrangements.

Solid floors

Solid floors with bedding or floor heat result in fewer drafts at the floor level than slotted floors. Clean farrowing pens or stalls on solid floors daily. Manure without bedding can be washed to a holding pit or lagoon. Do not use bedding with a liquid manure system because it clogs pumps and decomposes very slowly in pits and lagoons. See section on manure management.

Slope solid floors to alleys and drains. Use a two-slope floor if waterers are installed at the front of stall or pen, Fig 3.

Alley Slopes
½″/ft crown or side slope
⅛″/ft to drains

Floor Slopes
¼″-½″/ft, without bedding
¼″/ft, with bedding

Alley Widths
Feeding alley, 4′ for feed cart
Sow handling alley, 2½′
Remodeled buildings, as little as 2′ for feeding or sow alley

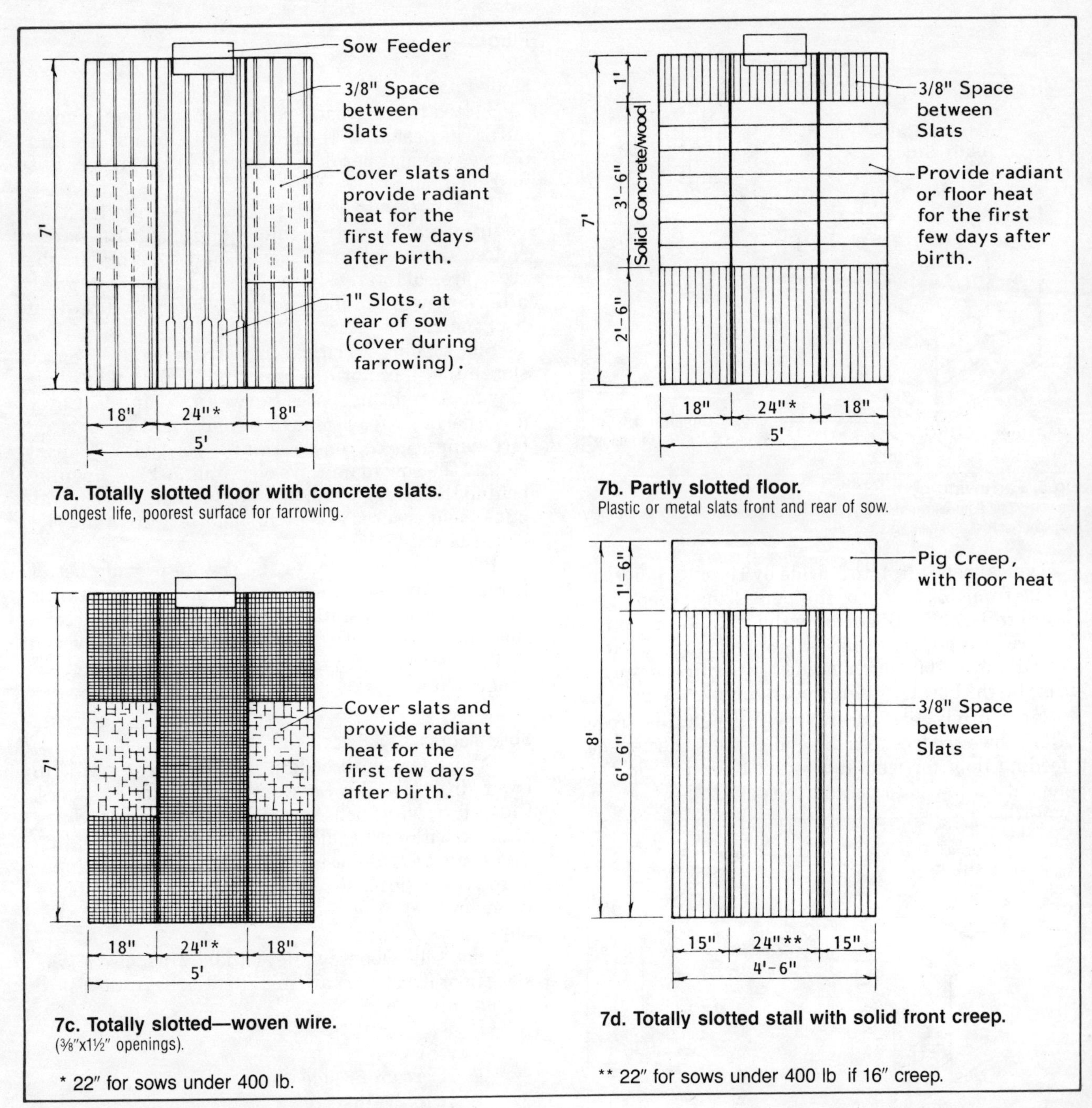

Fig 7. Slotted floor arrangements for farrowing.

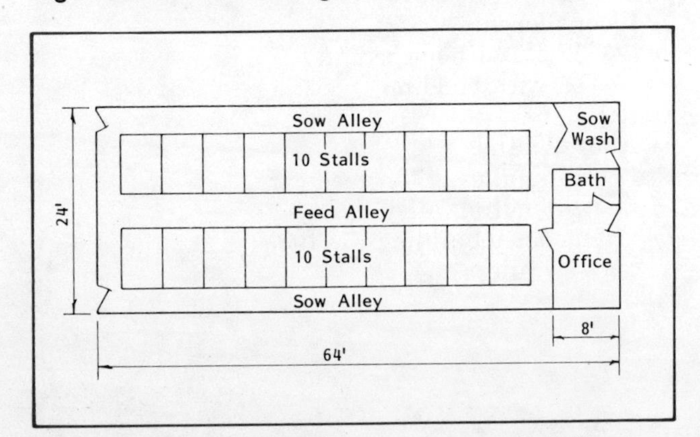

Fig 8. Farrowing building with two rows of stalls.
mwps-72680.

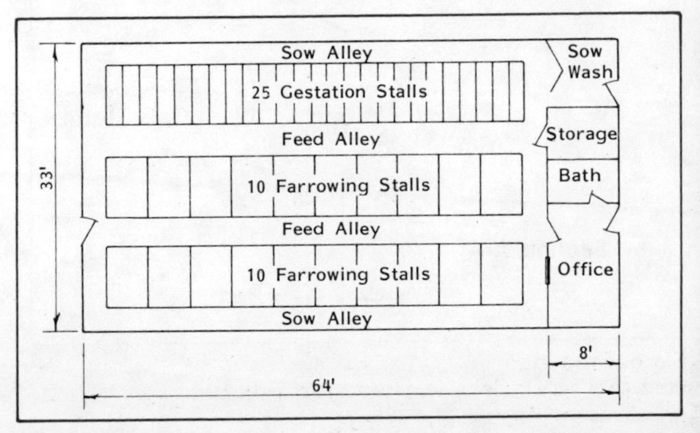

Fig 9. 20-sow farrowing building with 25 gestation stalls.
mwps-72699.

Floor Plans

Farrowing buildings commonly have 2 rows of farrowing stalls facing the center alley for feeding ease and sow interaction, Fig 8. Hand feed sows from a feed cart or feeding stations.

Fig 9 shows sow gestation stalls in the same room as farrowing stalls. Some producers hold sows in the farrowing building for 6 to 7 weeks before farrowing to encourage pig immunity to scours. This building will rarely be empty for total room cleaning.

Ten farrowing stall floor arrangements are shown in Fig 10. Fig 10a shows a two-slope solid floor, which often contains floor heat. Sows face the outside wall in this arrangement. Turn sows out twice a day for feed and water, to increase sow interaction, and to reduce cleaning of solid floors. A mechanical gutter cleaner just outside the rear of the stall can facilitate cleaning.

Fig 10b, 10c, and 10d show floors with slats at the rear of the stall. The concrete portion of the creep area often contains heat. Slope the front portion of the floor forward to drain spilled water. Dunging at the front of the stalls is a problem with pigs older than 3 weeks.

The cantilever floor allows frequent removal of manure, Fig 10e. Hand scrape manure from the gutter directly outside or into a deep narrow collection gutter. Scrape daily or weekly. The bottom of the gutter is level. A 1″ to 2″ lip at the end of the gutter retains liquids and reduces adhesion of solids to the gutter.

Fig 10f shows a floor with about 12″ slotted in the front and 30″ slotted over a gravity drain gutter at the rear.

Fig 10g and 10h show totally slotted floors under the stalls or the entire building. Fig 10i shows a raised partly or totally slotted floor stall with sows facing out. Scrape or flush manure toward center alley.

Fig 10j shows a two-row, solid floor remodeled unit. Slope the entire stall floor to the alley. Feed and water sows outside to increase interaction and reduce dunging in the stalls.

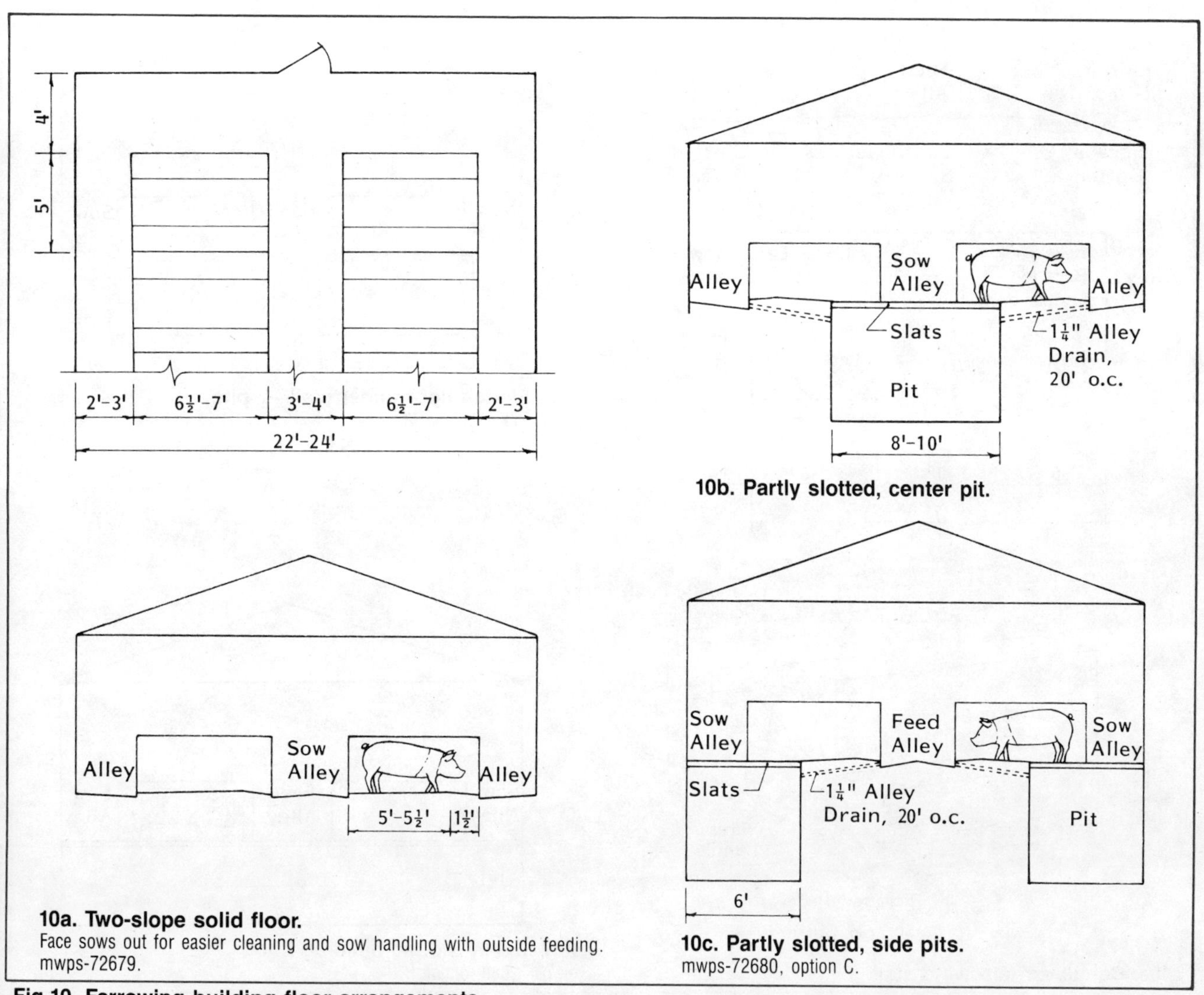

10b. Partly slotted, center pit.

10a. Two-slope solid floor.
Face sows out for easier cleaning and sow handling with outside feeding. mwps-72679.

10c. Partly slotted, side pits.
mwps-72680, option C.

Fig 10. Farrowing building floor arrangements.

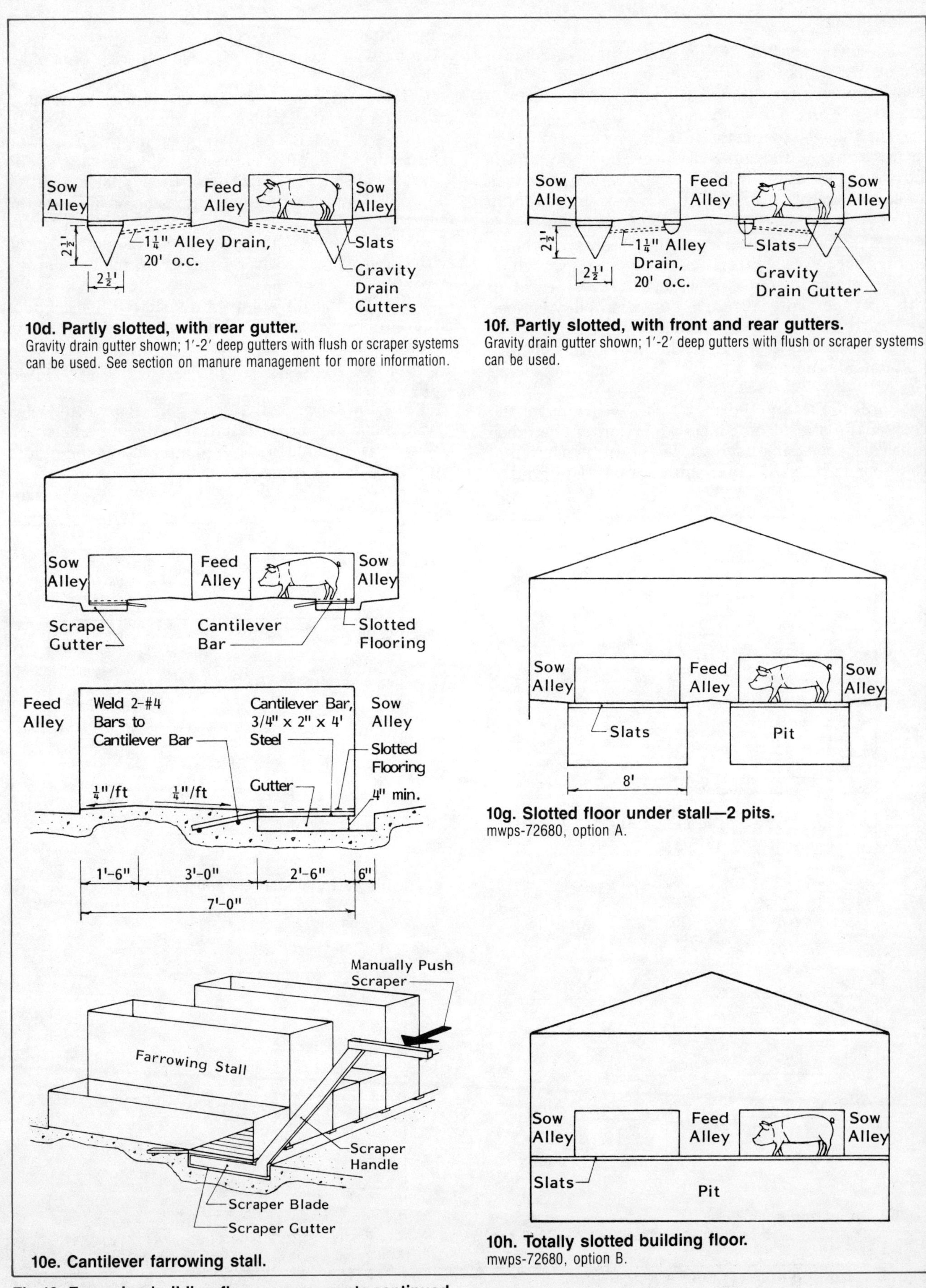

10d. Partly slotted, with rear gutter.
Gravity drain gutter shown; 1′-2′ deep gutters with flush or scraper systems can be used. See section on manure management for more information.

10f. Partly slotted, with front and rear gutters.
Gravity drain gutter shown; 1′-2′ deep gutters with flush or scraper systems can be used.

10e. Cantilever farrowing stall.

10g. Slotted floor under stall—2 pits.
mwps-72680, option A.

10h. Totally slotted building floor.
mwps-72680, option B.

Fig 10. Farrowing building floor arrangements continued.

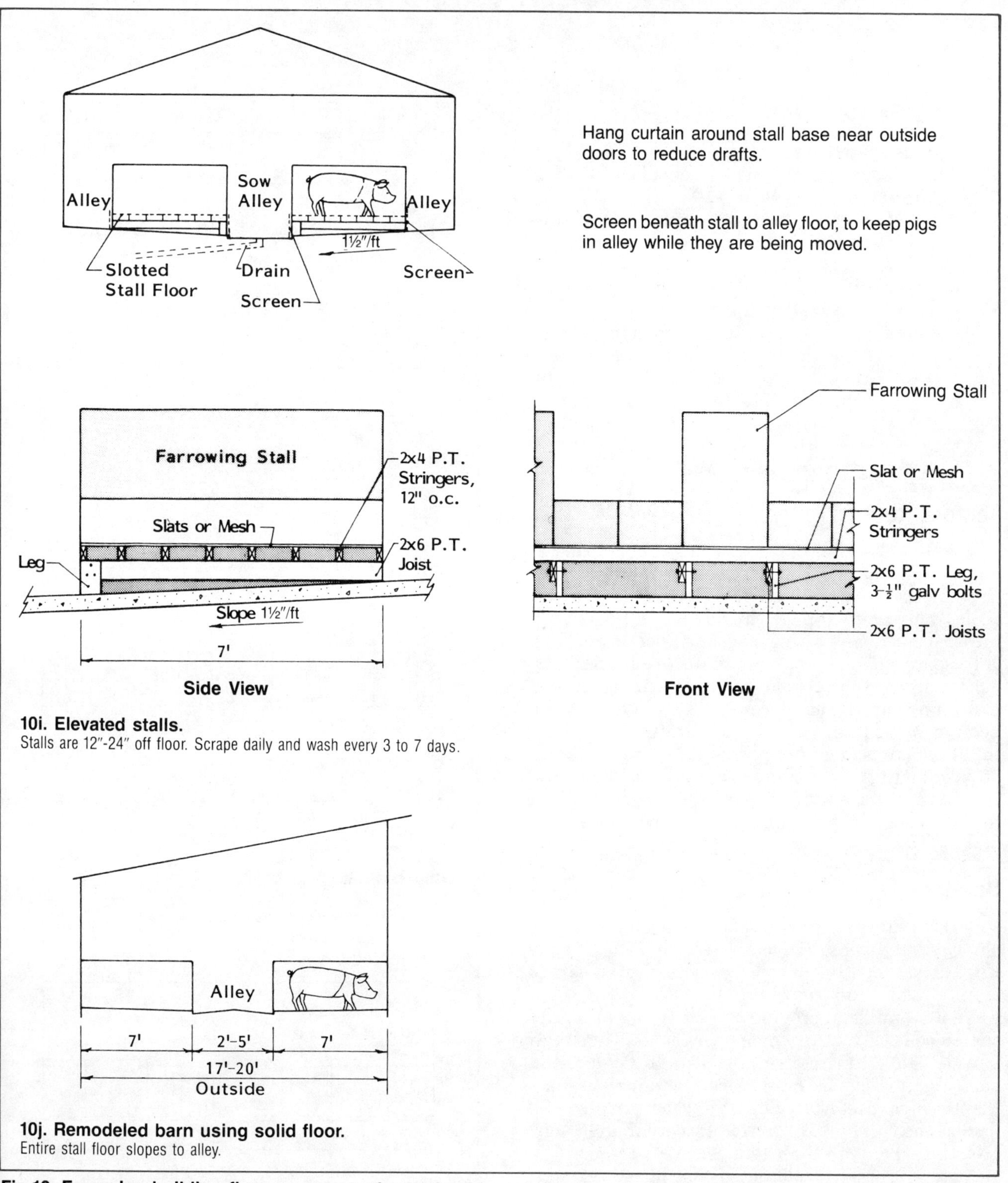

Hang curtain around stall base near outside doors to reduce drafts.

Screen beneath stall to alley floor, to keep pigs in alley while they are being moved.

10i. Elevated stalls.
Stalls are 12"-24" off floor. Scrape daily and wash every 3 to 7 days.

10j. Remodeled barn using solid floor.
Entire stall floor slopes to alley.

Fig 10. Farrowing building floor arrangements continued.

NURSERY FACILITIES

A nursery is for weaned pigs or for sows with litters. There are three types:

- **A late-wean nursery (pig-nursery)** for raising pigs weaned at 5 to 8 weeks from about 30 lb up to 75 lb.
- **An early-wean nursery (prenursery)** for raising pigs weaned at 3 to 4 weeks from about 12 lb up to 25-30 lb. After 3 to 4 weeks, move pigs to a pig-nursery.
- **A sow-pig nursery** for relocating sows and their 3-day to 3-week-old litters from the farrowing unit. After the sows are taken out, the pigs stay until they are moved at about 30 lb to a pig-nursery or at about 75 lb to a growing unit. Labor for moving sows with pigs is a disadvantage.

See building layouts in Fig 11-15.

Temperature

For 3-week-old pigs, provide 85 F temperature at pig level for the first few days after weaning. Lower the temperature 3 F per week to a minimum of about 70 F for 8-week-old pigs. Provide warm floors with infrared heaters, heating pads, or floor heat. With only space heaters, you may have to set the thermostat higher than the desired room temperature to maintain warm floors. For prenursery, consider preheating ventilation air to reduce drafts.

In a non-bedded sow-pig nursery, two different environmental conditions are required. Keep the room temperature at 65-75 F with supplemental heat in the creeps. Room temperature can be 60 F if bedding is used.

Space

In a pig-nursery, provide 3 to 4 ft^2/pig for totally or partly slotted pens. Size each pen for 16 to 20 pigs. Keep litters together as much as possible, but sort to maintain size uniformity.

In a prenursery, provide 2 to 2½ ft^2/pig in totally slotted pens with woven wire floors or slats less than 2″ wide. Avoid concrete slats, slats over 2″ wide, and partly slotted pens. These space allowances assume that the pigs enter at about 12 lb and leave at about 30 lb. Size each prenursery pen for no more than 16 pigs; preferably no more than 10 pigs.

In totally or partly slotted sow-pig nurseries, minimum pen size is 5′x10′ for one sow and litter; 8′x10′ for two. Minimum pen width is 5′ and minimum pen length is 10′. Provide 1 ft^2 of heated creep area per pig in each pen. Allow an additional 40 ft^2 of open lot per sow and litter in shed and lot systems. Do not combine more than two sows and litters per pen.

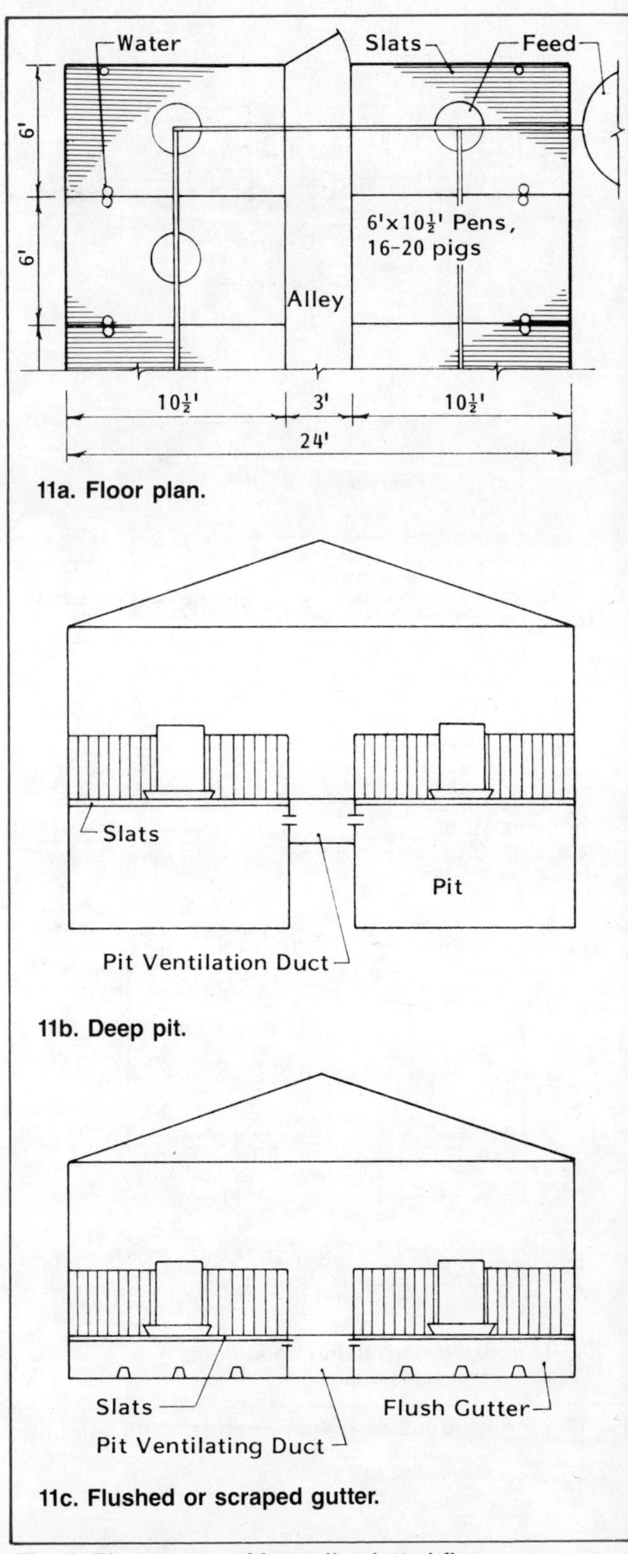

11a. Floor plan.

11b. Deep pit.

11c. Flushed or scraped gutter.

Fig 11. Pig-nursery with totally slotted floor.
mwps-72686.

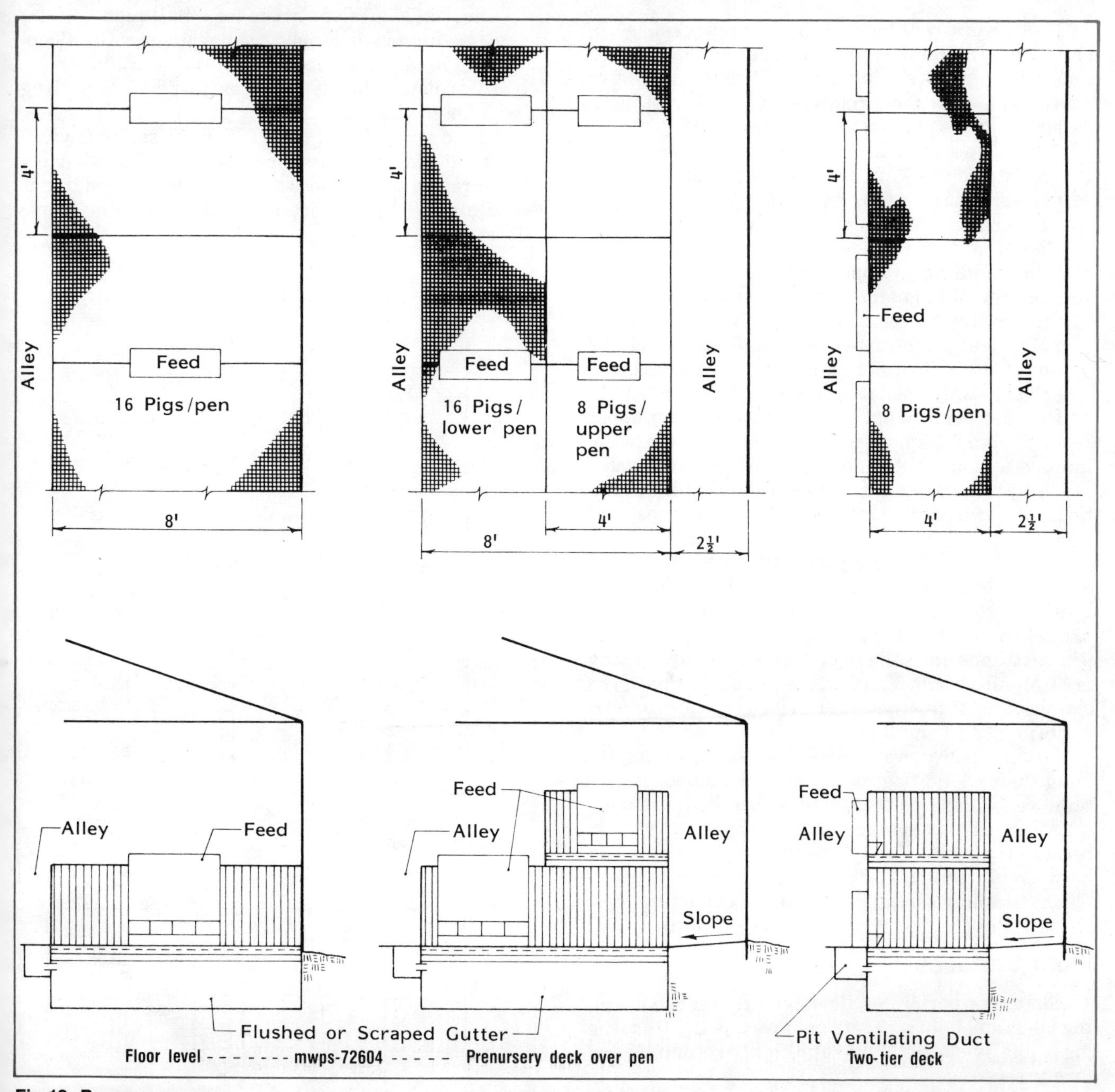

Fig 12. Prenursery pens.
Floors and decks are ⅜"x1¼" woven wire mesh.

Facilities

Use fenceline feeders in prenursery pens and decks, and in partly slotted pig-nurseries. Round or fenceline feeders can be used in totally slotted pig-nurseries. Provide adequate feeder space:

Pig weight, lb	Feeder space
12-30	2 pigs/space
30-50	3 pigs/space
50-75	4 pigs/space

Section the openings of feeders for early weaned pigs so the pigs cannot crawl into them. Leave at least 6" between feeder and pen partitions to prevent dunging in end feeder holes.

Provide one waterer for each 10 prenursery or pig-nursery pigs, with a minimum of 2 waterers per pen. Place nipple height at about 10" for 10 lb pigs; about 12" for 12 lb. Raise accordingly as pig grows.

Make pen partitions 24" high in prenurseries and 32" high in pig-nurseries.

Decks

Decks work well for prenursery pigs because improved environment and increased operator attention reduce postweaning stress. Decking increases nursery capacity for a relatively small investment, because pigs are placed in a previously unused space.

Advantages include:

- Increased stocking density per building.
- Lower operating costs per pig.
- Fewer pigs per group in smaller pens.

Disadvantages include:

- Pig handling in upper decks is difficult.
- Design and management of environmental control system is more critical.
- Maintenance of upper deck equipment may be more difficult if not arranged for easy access.
- Pigs in lower pens can be dirtier than normal.

Pigs are usually put in and removed from upper decks by hand. Remove pigs before they reach 50 lb, limit deck depth to 4′, and use removable front gates to ease pig handling. Arrange feeders, solid partitions, and overlays in both decks to facilitate observation.

Consider the higher pig density and the influence of the top decks on airflow patterns when selecting a ventilating system. Never locate pigs in the direct path of the ventilating air. Reduce drafts with solid pen partitions and solid floor overlays in the sleeping area. Monitor room temperature and airflow patterns closely because fluctuations can lead to scours or poor performance in small pigs.

Locate the waterer at rear of pens to train pigs to dung there. Solid floor overlays near the feeder also improve dunging patterns and reduce feed wastage.

All-in, all-out management reduces disease risk, because it allows thorough cleaning and disinfecting between pig groups. Keep facilities as clean as possible, but avoid washing pens when occupied by pigs. Dry scraping is best when pigs are present.

Floor Surfaces

Slotted or perforated floors greatly reduce cleaning labor and help separate manure quickly from the pigs. Totally slotted floors are highly recommended for nursery pigs.

For partly slotted floors, slot at least two-thirds of the floor in prenurseries and at least 40% in pig-nurseries. It is difficult to train nursery pigs to sleep on solid floors and dung on slats, so consider the following:

- Use a long, narrow pen shape (length 2 to 4 times the width). Minimum pen width is 4′ for prenursery (5′ for pignursery) with partly slotted floor.
- Use solid partitions over solid floors and open partitions over slotted areas. In buildings with ceiling slot air inlets, use open partitions along alleys for cross ventilation in summer.
- Place feeders on solid floors and waterers over slats.
- Provide a 1″-2″ step up from slotted floors to solid floors to separate dunging and sleeping areas and to keep manure off the solid floor.
- During cold weather, provide supplemental zone heat in solid floor areas for small pigs.
- Use hinged or removable hovers over the sleeping areas to reduce drafts and retain heat.
- Wet down the slotted area just before putting pigs in the pen.
- Feed on the solid floor for the first few days.

Avoid concrete slats for pigs under 30 lb—select nonabrasive slats. Slope solid floors where bedding is used about ¼″/ft for drainage. If no bedding, slope floors ½″/ft.

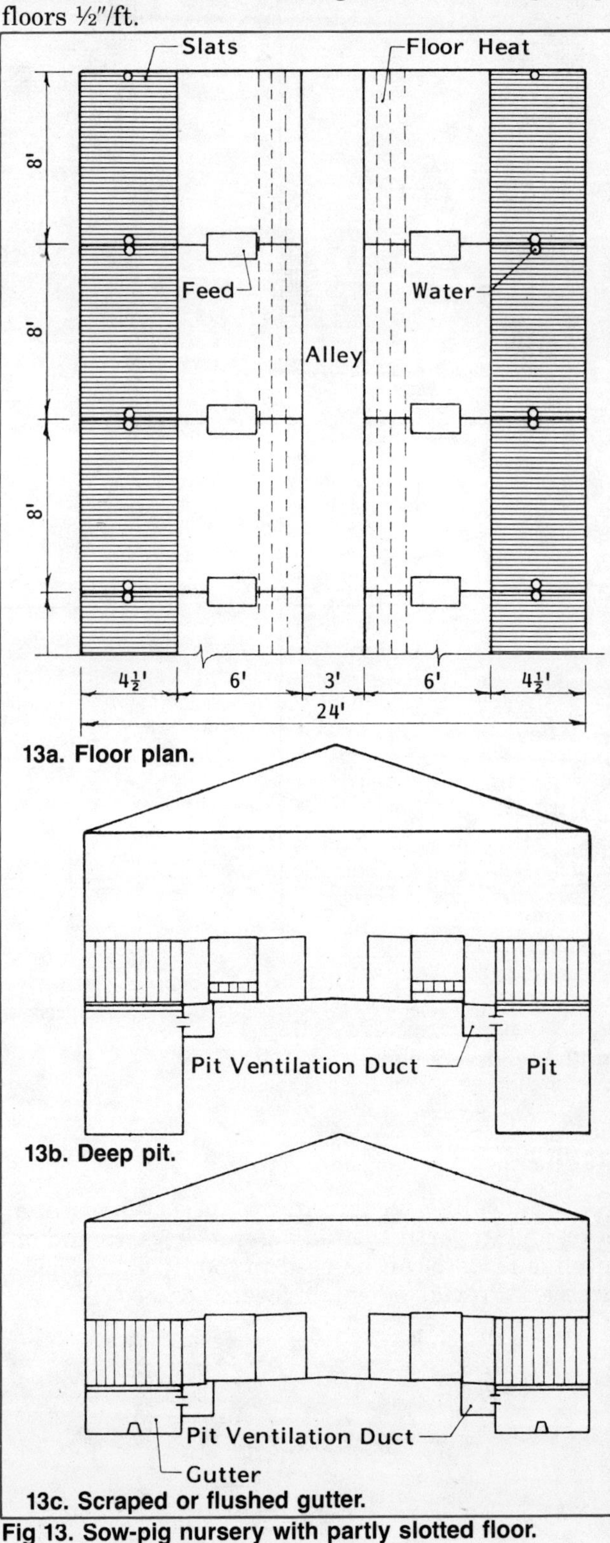

13a. Floor plan.

13b. Deep pit.

13c. Scraped or flushed gutter.

Fig 13. Sow-pig nursery with partly slotted floor.
Two sows and litters per pen. mwps-72685.

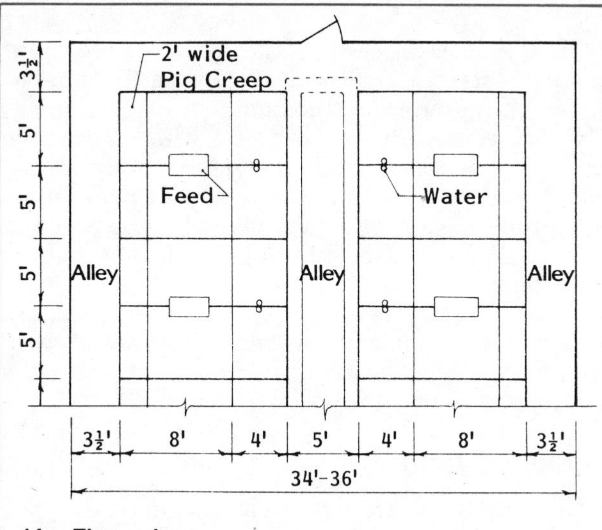

14a. Floor plan.

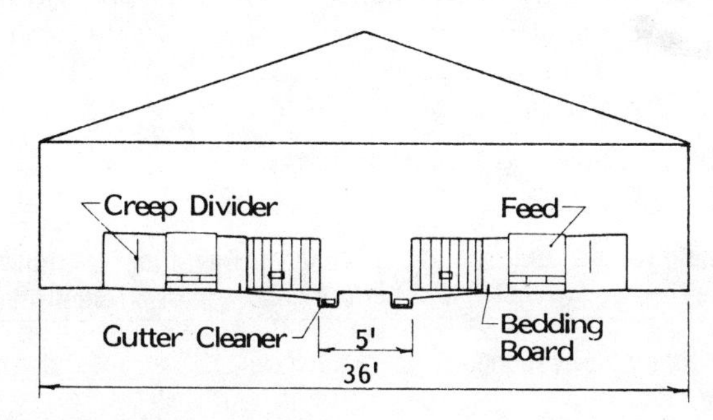

14b. Barn cleaner in alley.

Periodically scrape manure into gutter. Alley can be flat and scraped with small tractor.

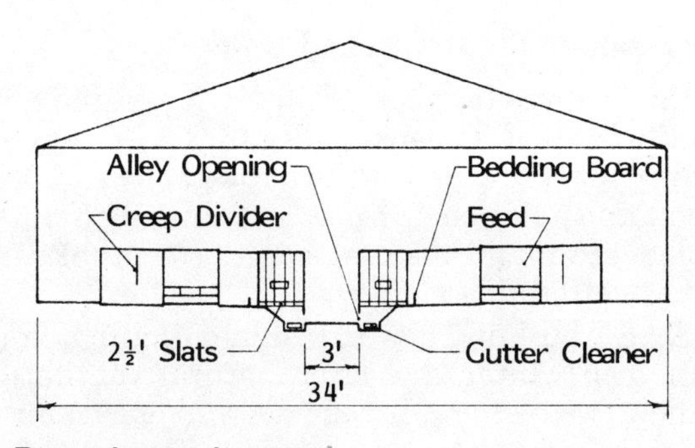

14c. Barn cleaner in pens.

Alley sides are solid except provide one 6"x18" scrape opening per pen.

Fig 14. Sow-pig nursery with bedded pens and barn cleaner.

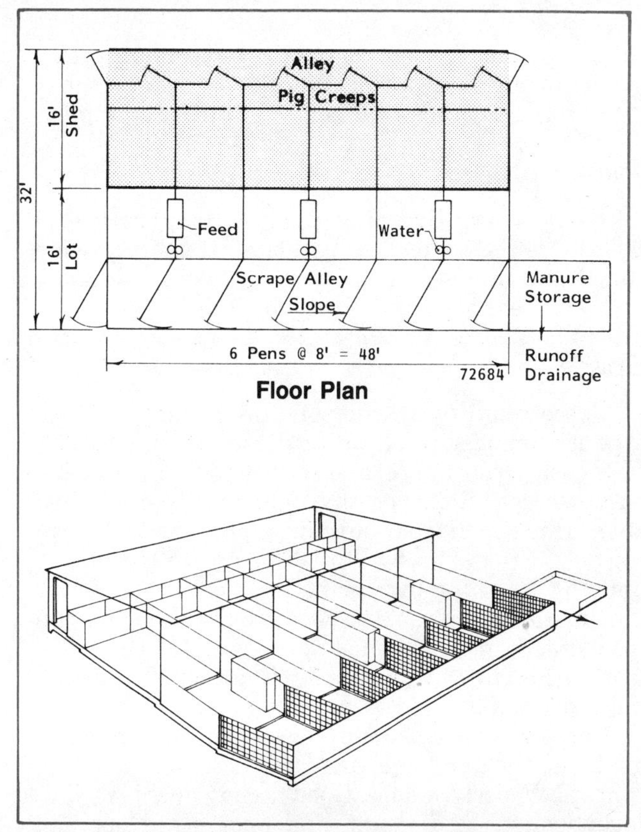

Fig 15. Sow-pig nursery with tractor-scraped open lot.

Face open front south or east. mwps-72684.

GROWING AND FINISHING FACILITIES

In this handbook the growth stages of pigs are:

- Growing stage - 75 to 150 lb.
- Finishing stage - 150 to 220 lb or market weight.

Temperature

For growing-finishing pigs, a temperature of 60-70 F is recommended. Although finishing swine can thrive in lower temperatures, they grow faster and with less feed in the recommended range.

Space

Space requirements depend on animal size and type of housing. Overcrowding causes animal discomfort, poor performance, and increases stress and disease susceptibility. For environmentally controlled or modified open-front buildings, provide 6 ft^2/growing pig (75 to 150 lb), 8 ft^2/finishing pig (150 to 200 lb), and 9 ft^2/finishing pig during hot weather or if over 200 lb. As pigs grow, move each group to a larger pen to maintain recommended space per pig. For open-front sheds with lot, provide 5-6 ft^2/growing-finishing pig indoors and 12-15 ft^2/pig outdoors.

Pen size and pigs/pen vary with management goals. As the number of pigs/pen increases, competition affects performance. Good management, including adequate feed, water, and floor space per pig, reduces the adverse effect of large groups. A reasonable maximum is 20 to 25 pigs/pen (3 litters) in buildings with partly or totally slotted floors.

Facilities

Provide one drinking space for each 15 pigs with a minimum of 2 waterers per pen. Place permanently mounted nipple waterers at 26″ and install a 4″ step for smaller pigs. Adjust movable nipple waterers at about the height of the pig's back.

Use fenceline feeders on partly slotted floors and round or fenceline feeders on totally slotted floors. Provide one feeder space per 4 to 5 pigs.

In buildings with partly slotted floors, use solid partitions over solid floors and gated partitions over slotted areas to promote good dunging habits. Use hog-proof materials on walls and partitions exposed to swine. Concrete partitions are recommended on solid floors. Over slats, consider vertical pipe spaced 6″ o.c. to prevent pigs from climbing. Make pen partitions at least 32″ high.

Floor Surfaces

Slope solid floors to facilitate manure handling and cleanliness:

- ½″/ft for solid or partly slotted floors.
- ¼″/ft for bedded solid floors.
- ½″/ft for outside feeding floors.
- ¼″/ft for inside feeding floors.
- ⅛″/ft to drains for alleys.
- ½″/ft crown or side slope for alleys.

Make at least 30%-40% of the pen floor slotted. For concrete slats, use 5″-8″ wide slats with 1″ slots. See section on manure management.

Partly slotted floors work best when the pen is about 20′ long (8′-10′ slotted) and 6′-10′ wide. Partly slotted buildings with a single row of pens (alley along front or back) generally result in good dunging patterns. Partly slotted buildings, with pens on both sides (center or off-center alley), require good management to train pigs to dung on the slotted area. See nursery section for other design factors which help toilet train pigs on partly slotted floors. Pen shape is not important on totally slotted floors.

Building Types

Growing-finishing pigs are housed in three types of buildings, Table 2:

- Environmentally controlled—fan ventilated in winter; often naturally ventilated in summer.
- Modified open-front—naturally ventilated.
- Open-front with outside lot.

Table 2. Finishing building types.
Summer performance of the three types is about the same. Approximate rank.

Building type	Initial cost	Winter performance	Operating cost	Labor requirements
Environmentally controlled	Higher	Higher	Higher	Lower
Modified open-front	Lower to medium	Higher	Lower	Lower
Open-front/ outside lot	Lower	Lower	Lower	Higher

Environmentally Controlled Units

Environmentally controlled (EC) buildings are usually the most expensive, Fig 16. An EC building is usually 32′-44′ wide with two rows of pens. They are often totally slotted, because bad dunging habits and inadequate manure storage can be a problem with partly slotted floors.

An EC building has an insulated ceiling, fan ventilation in winter, and often natural ventilation in summer with sidewall vent doors. Vent doors must seal tightly for effective winter ventilation. Odor levels in EC buildings with pit manure storage are high but can be reduced with pit ventilation. See section on ventilation.

Modified Open-Front Building

A modified open-front (MOF) building is insulated and naturally ventilated, Fig 17. The building is usually 28′-33′ wide with a single row of pens. The floor can be partly slotted or have an open gutter with a flush or scrape system. Floors can be totally slotted in mild climates where building temperatures stay above 60 F. Make partly slotted floors about 30%-40% slotted.

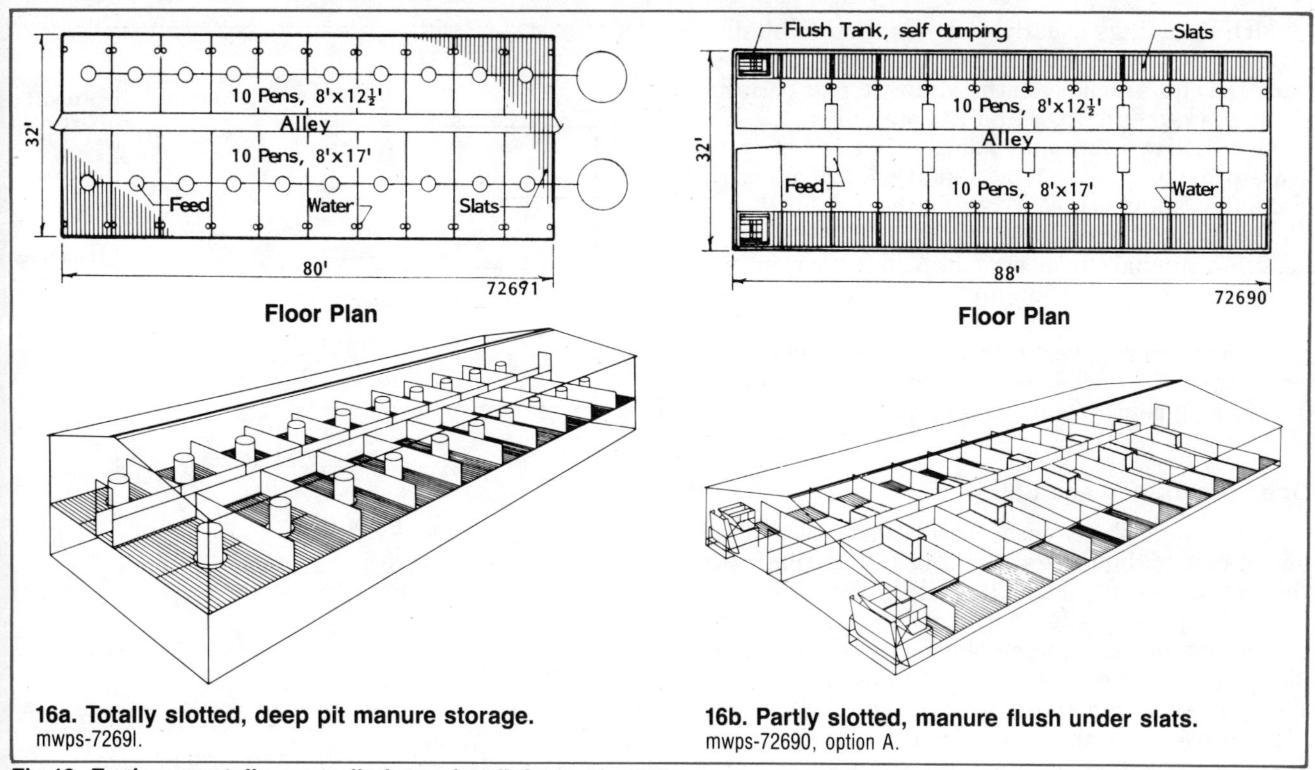

16a. Totally slotted, deep pit manure storage.
mwps-72691.

16b. Partly slotted, manure flush under slats.
mwps-72690, option A.

Fig 16. Environmentally controlled growing-finishing buildings.
Off-center alleys and two pen sizes allow you to move growing pigs across alley to finishing pens.

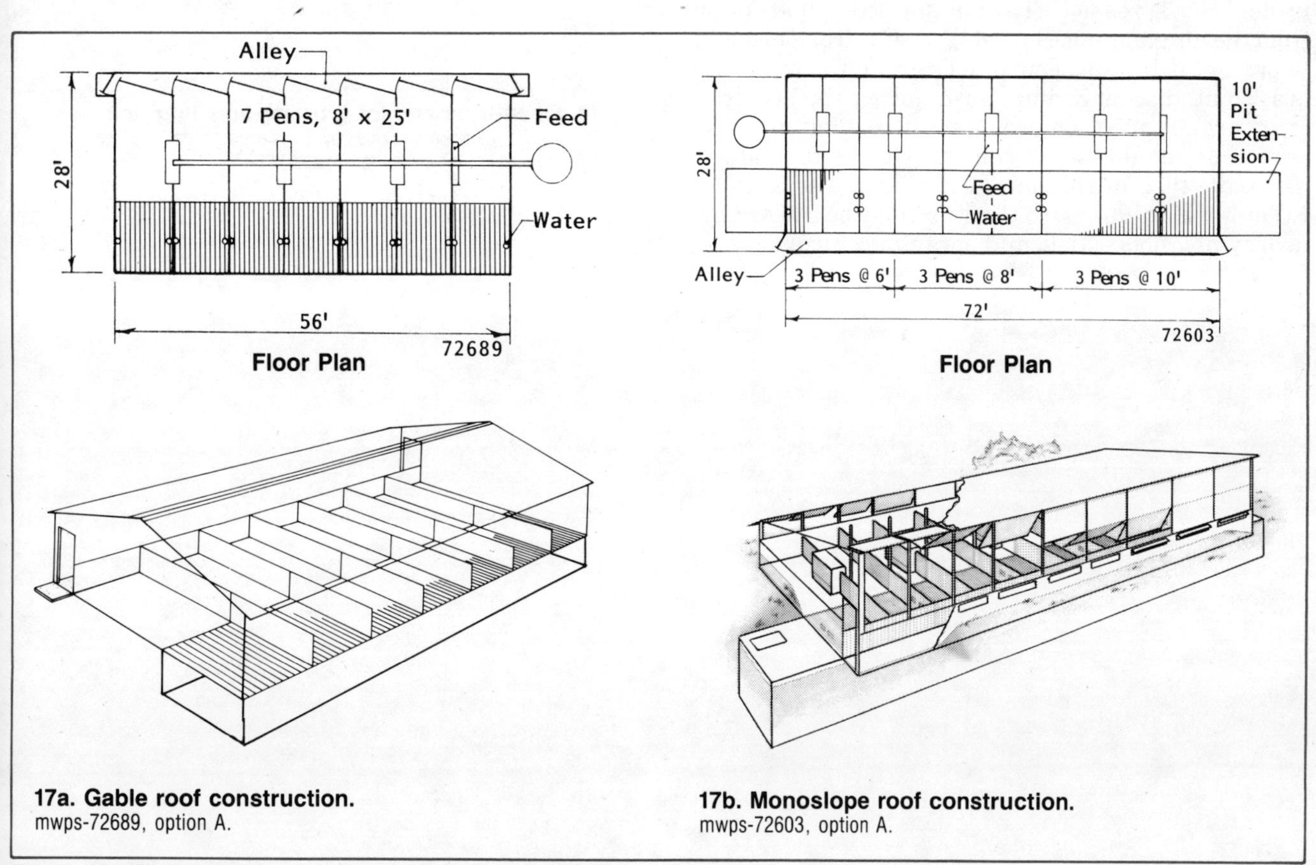

17a. Gable roof construction.
mwps-72689, option A.

17b. Monoslope roof construction.
mwps-72603, option A.

Fig 17. Modified open-front, growing-finishing buildings.
These buildings are partly slotted with long narrow pens. Deep pits and flush or scrape gutters can be used under the slats.

MOF buildings usually cost less than EC buildings of the same capacity. They use less energy because they have no fans. In the Midwest, pigs perform about the same in MOF and EC buildings.

Open ridge vents and adjustable sidewall vent doors provide ventilation. Adjust vent doors with manual or automatic controls. For best natural ventilation, orient the building perpendicular to prevailing summer winds. In most of the Midwest, orient the long axis east-west. Adequate insulation reduces winter condensation and summer radiation heat gain, and conserves heat to maintain suitable winter temperatures. Provide overhead or floor zone heat in the growing pens to start pigs moved in from a nursery.

Open-Front Shed and Lot

Open-front sheds with outside lots are usually lowest cost of the three types, Fig 18. Ventilation is through the open front and doors in the back wall. The ceiling is insulated to control winter condensation and provide summer shade. Construction requires less skill because most of the concrete is in flat slabs rather than manure pits. The smaller growing pigs do not perform as well in this building type during cold weather, but may perform better in warm weather.

Outside lots require more labor than buildings with slats. Handle manure as a solid with a scraper, loader, and spreader. Channel surface runoff away from the unit and slope lots at ½"/ft. Control lot runoff to prevent pollution of surface or ground water. A lot is a "point" discharge and may require runoff control structures to meet environmental regulations. See section on manure management. Open lots produce more odor (because of their larger surface area) than EC or MOF buildings, so their use may be limited by nearby neighbors. Bait and spray to control flies.

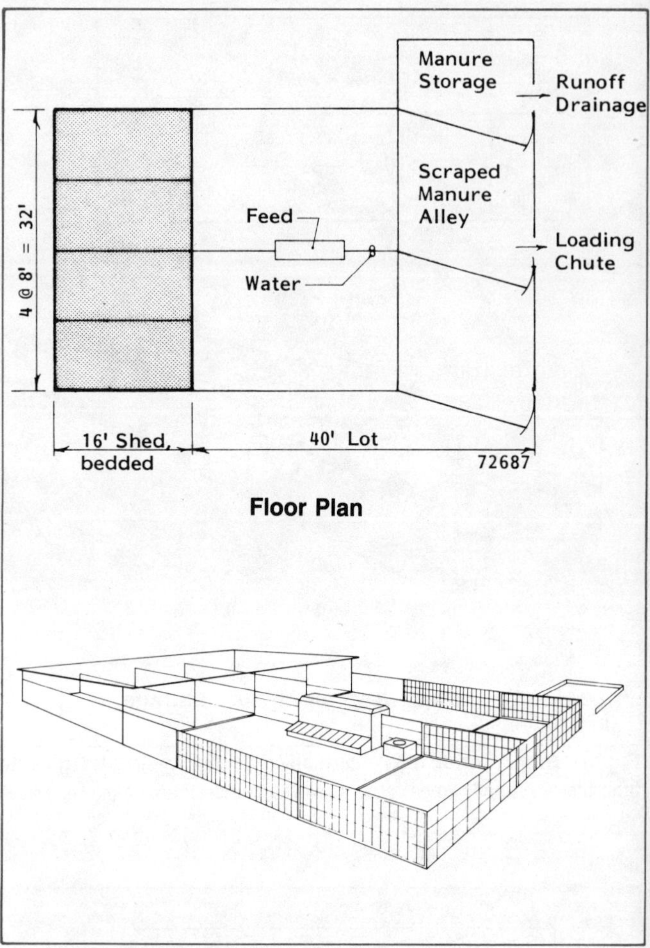

Fig 18. Open-front and lot growing-finishing unit.
The 16′ open-front shed provides a covered sleeping area. Pigs feed, water, and dung outside on the paved lot. mwps-72687.

GESTATION AND BREEDING FACILITIES

Alternatives for gestation and breeding facilities range from pasture systems with limited shelter to environmentally controlled facilities with gestation stalls or tethers.

Advantages of environmentally controlled (EC) housing for breeding-gestation include:

- Better control of mud, dust, and manure.
- Reduced labor for feeding and sow handling.
- Better supervision of breeding program.
- Improved control of internal and external parasites.
- More control over sow feed intake.
- Smaller land requirements.
- More efficient use of boars.
- Improved operator and animal comfort and convenience.

Disadvantages include:

- Higher initial investment.
- Possible delayed sexual maturity, lower conception rates in gilts, and lower rebreeding efficiency in sows with average management.
- Requires more intensive management and daily attention to details.

Facility Sizing

Provide space for gestating sows and gilts, replacement gilts, recently weaned sows, boars, and breeding pens.

In an intensive breeding program, overbreed to ensure that buildings are kept full. Some intensive farrowing schedules require that sows be bred on the first heat cycle after farrowing. With an 80% conception rate, 20% of the sows do not conceive and are sold and replaced with bred gilts.

For some farrowing schedules, sows that do not conceive after the first heat are added to the following group of sows. Sows open after the second heat cycle are culled.

Also, some low performance sows are culled from the farrowing group after weaning. With all factors considered, provide 3 to 6 times as many gilts in the gilt pool as will be needed for replacement at breeding. The larger number is for hot weather breeding when conception rates are lower.

The required space for boars and gestating females depends on the farrowing schedule. See scheduling section for estimating the size of breeding-gestation facilities.

Open-Front Shed With Lot

An open-front shed with lot provides a well-bedded sleeping area protected from adverse weather and an outside lot for feeding, watering, and dunging. This system moves the breeding-gestation operation off pasture and out of the mud with low-investment or remodeled facilities. Odor and flies are a problem because manure is deposited over a large area and remains exposed for a long time. Snow tends to drift into the lot, making feeding inconvenient and allowing animals to cross fences. Properly placed snowfence reduces snow accumulation in the lot. A solid windbreak along the outside of the lot provides wind protection but may increase snowdrop.

Fig 19 shows typical housing for sows from a 20-sow farrowing building with 6 or 7 farrowings per

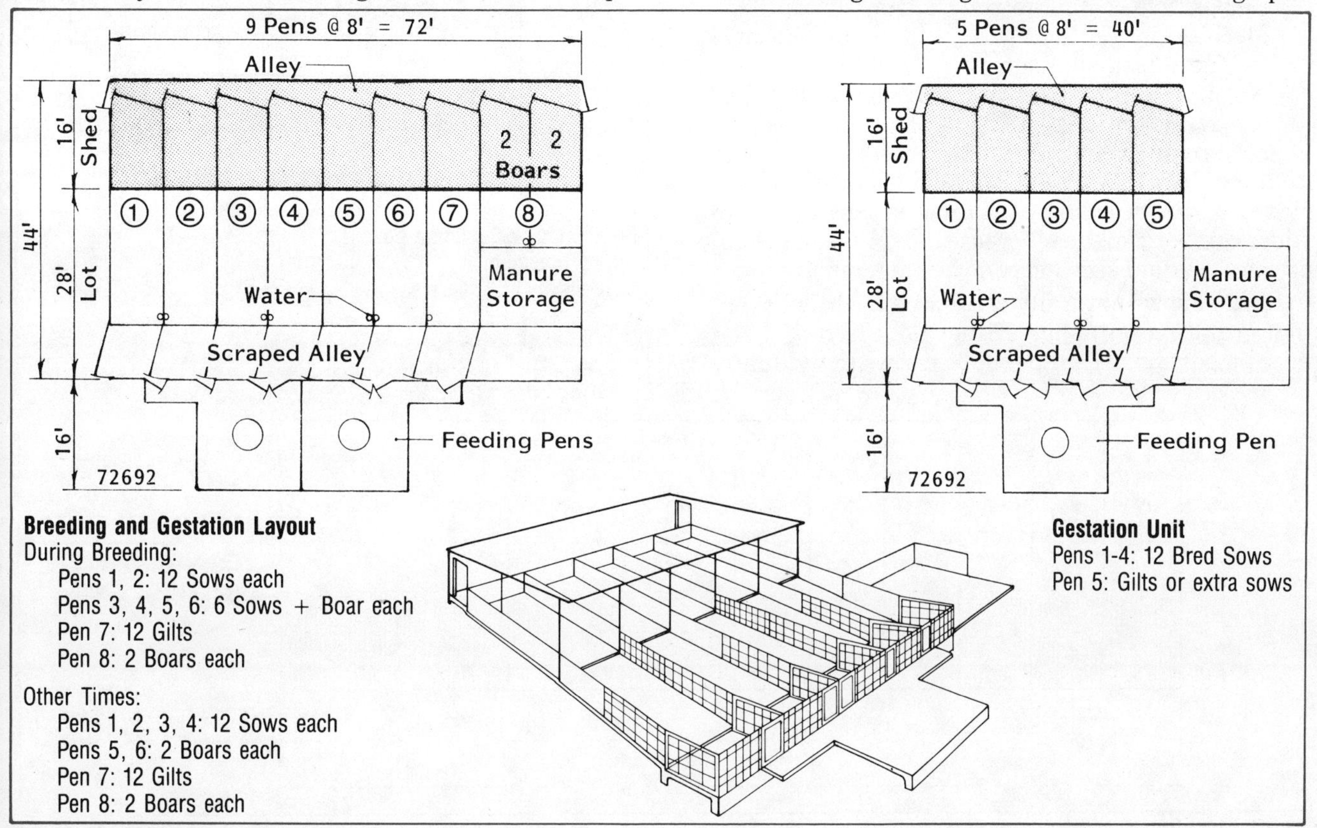

Fig 19. Open-front and lot gestation unit.
mwps-72692.

year (3 sow groups). Allow about 8 ft² of housing area and about 14 ft² of lot area per sow in gestation pens. Provide 40 ft² of housing area and 40 ft² of lot area per boar in individual or group pens. Double these areas in the breeding pens. Group up to 12 sows per gestation pen and no more than 6 sows (plus boar) per breeding pen.

Environmental control for a shed is limited. Use solid partitions between pens in the shed and use doors covering the top half of the open front to reduce drafts. Leave a 2″ crack above the door for escape of warm, moist air. During cold weather, provide plenty of dry bedding in the shed.

Handle manure scraped from the lot as a solid. Provide a manure storage area and a conventional manure spreader. Channel lot runoff to an approved management system, such as a settling basin or infiltration channel. See section on manure management.

Limit-feed gestating sows to control excessive weight gain and to prevent feed wastage. Feed handling options include:

- Daily floor feeding in the pen.
- Self-feeding on a bulky diet.
- Turnout interval self-feeding.
- Daily individual feeding in closable stalls.

For floor feeding, distribute the feed so all sows have access without competition. Floor and interval feeding require the least labor.

Enclosed Sow Housing

Options for enclosed sow housing include:

- Mechanical ventilation year-round; often with evaporative cooling to relieve heat stress.
- Mechanical ventilation in winter and natural ventilation in summer.
- Natural ventilation year-round in modified open-front buildings.

Space requirements are listed in Table 3. Group **no more** than 12 sows per pen and **no more** than 6 (plus boar) in breeding pens. Sow groups of 5 or 6 reduce competition and increase flexibility in grouping sows of equal size and compatible temperament.

Fig 20 shows naturally ventilated modified open-front gestation buildings. Either floor feed daily on the solid portion of the floor or feed sows individually

Table 3. Space requirements of housed breeding swine.

Animal type	Weight lb	Group pens: Solid floor ft²	Group pens: All or partly slotted floor* ft²	Group pens: Animals per pen	Stall size
Breeding					
Gilts	250-300	40	24	up to 6	
Sows	300-500	48	30	up to 6	
Boars	300-500	60	40	1	2′-4″x7′
Gestating					
Gilts	250-300	20	14	6-12	1′-10″x6′
Sow	300-500	24	16	6-12	2′-0″x7′

* Use this column for flushed open gutter. Open gutter not recommended in breeding because of slick floors.

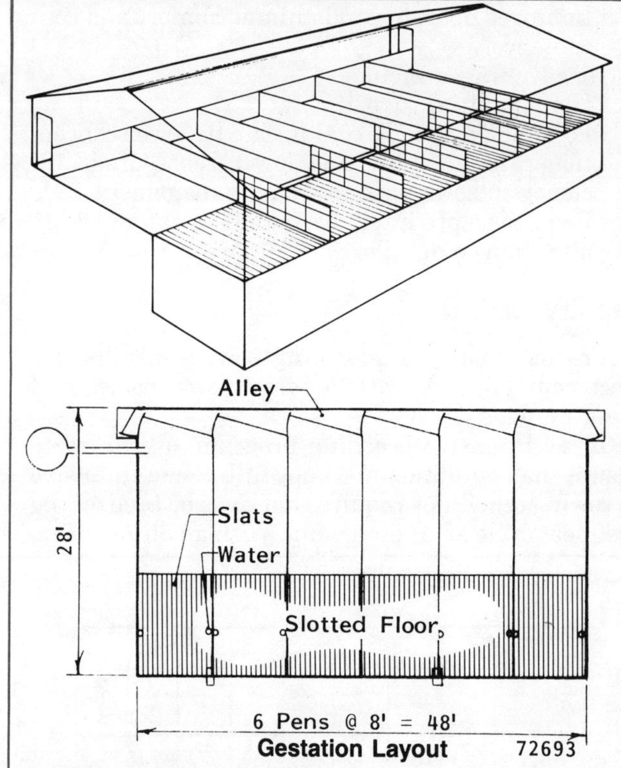

Breeding and Gestation Layout
During Breeding:
Pens 1, 3: 2 Boars each
Pen 2: 12 Gilts
Pen 4: 12 extra Sows
Pens 5, 6: 12 Bred or Unbred Sows
Pens 7-10: 1 Boar and 6 Sows or 12 Bred Sows

Fig 20. Modified open-front gestation unit.
mwps-72693, option A.

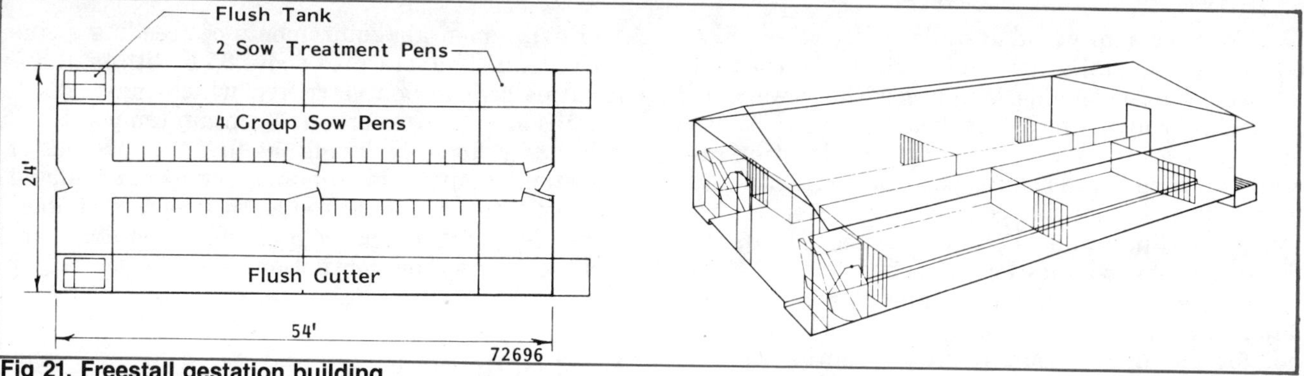

Fig 21. Freestall gestation building.
mwps-72696.

in freestalls along the side of the pen. Fig 21 shows a naturally ventilated gestation building with pens and freestalls for individual sow feeding.

Freestalls are either full size, 2′x6′-7′, or abbreviated 1½′x2′ deep. Arrange freestalls on partly slotted floors so sow manure is deposited on the slats. Body (or girth) and neck tether stalls cost less than conventional stalls, but some sows have difficulty adjusting to tethers. Place gilts in conventional stalls for a few days before placing in tether stalls.

Fig 22 shows mechanically ventilated gestation buildings with 2, 3, and 4 rows of stalls. Each sow is fed individually in her stall. These buildings can be naturally ventilated in summer.

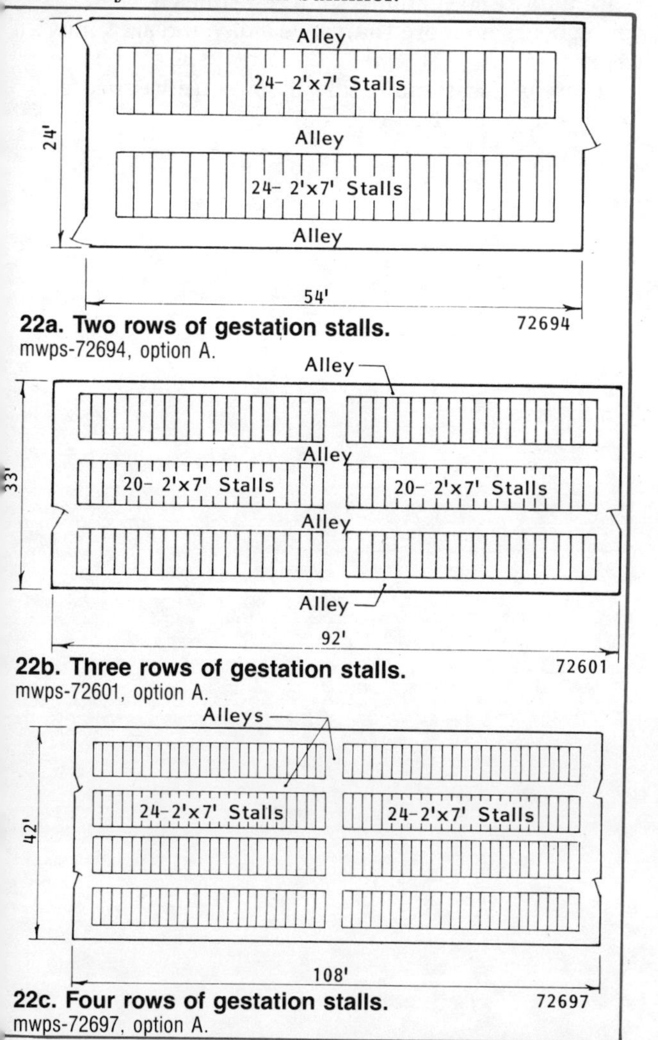

22a. Two rows of gestation stalls.
mwps-72694, option A.

22b. Three rows of gestation stalls.
mwps-72601, option A.

22c. Four rows of gestation stalls.
mwps-72697, option A.

Fig 22. Environmentally controlled gestation buildings.

Breeding Buildings

Breeding buildings are economically justified only for weekly or continuous farrowing schedules. Breeding buildings are environmentally controlled to eliminate hot and cold weather breeding slumps. See section on ventilation.

A breeding building has holding pens and/or stalls or tethers, boar pens and/or stalls, and breeding pens. Pen layout affects the social interaction between boars and females, a vital part of breeding performance. Alternate boar pens with female pens and use open pen partitions to encourage social contact. Locate boars as close as possible to recently weaned sows to induce strong heat indications. See Fig 23 for example floor plans of breeding buildings.

Developing and breeding gilts in environmentally controlled buildings presents special problems. Gilts are most fertile when housed in small groups with adequate floor space, Table 3. Exposure to boars hastens the onset of puberty, so put boars in adjacent pens.

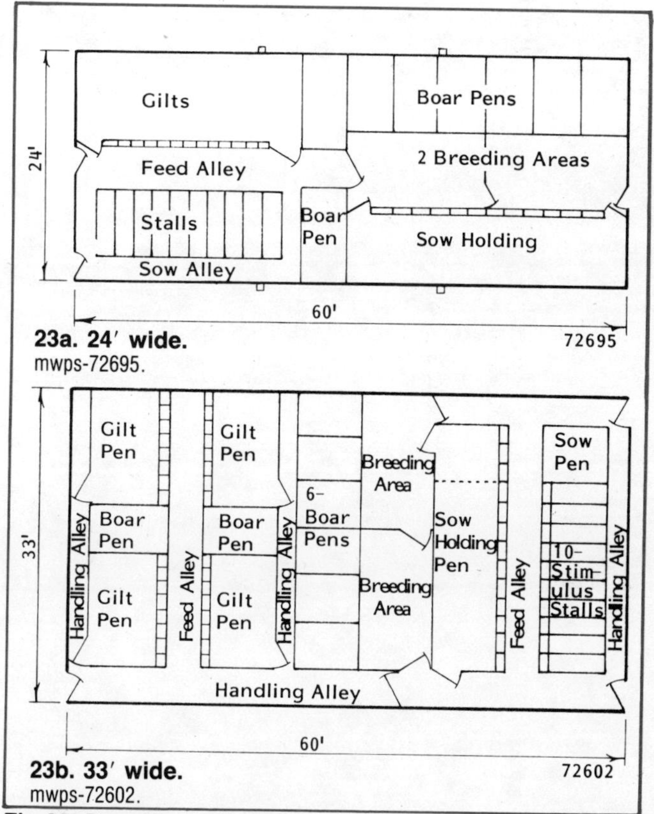

23a. 24′ wide.
mwps-72695.

23b. 33′ wide.
mwps-72602.

Fig 23. Breeding buildings.

Sows are usually held in stalls (called stress, dry-off, or stimulus stalls) to prevent fighting for the first 2 to 3 days after weaning. Weaning within 6 weeks of farrowing triggers the estrus (heat) cycle 4 to 8 days after pigs are removed. After holding in stimulus stalls, move sows to a holding pen where it is easier to detect heat. Locate this pen close to the boars, because fenceline contact between sows and boars induces a stronger indication of heat.

An alternative to stimulus stalls are pens holding 4 to 6 sows after weaning. Keep a mature boar in the pen for the first 48 hours to reduce fighting among newly weaned sows and to stimulate the onset of estrus. After 48 hours, move the boar to an adjacent pen for fenceline contact.

After heat is detected, place the sow in a breeding pen with the boar. The actual heat period (when the sow stands and allows the boar to mount) lasts about 2½ days. Breed 22 to 26 hr after the onset of standing heat (16 to 20 hr for gilts). After breeding, return the female to a holding pen; breed again 8 to 12 hr later. Breed in the boar pen or a separate neutral pen. A neutral pen provides better footing because the flooring has a chance to dry between uses. Slick floors result in poor breeding performance. Improve footing by roughening concrete slats.

See Table 3 for space requirements in breeding buildings. Make breeding pens at least 8'x8' to allow animals room to mount. Use pen partitions at least 42" high. Use vertical planks or bars to keep animals from "climbing" partitions. Arrange pens and alleys so you can see all animals from adjacent walkways.

Environmental requirements of breeding swine are critical. Temperatures above 85 F with high humidities reduce boar fertility, which may take 3 months or longer to correct. Maintain temperatures below 85 F for sows during the first 2 to 3 weeks of gestation to ensure maximum litter size and during the last 2 to 3 weeks to reduce stillborns and abortions. Temperatures below 55 F affect operator comfort. See section on ventilation.

Boar Housing

House boars in individual stalls or small individual pens. Grouping boars leads to fighting, homosexual activity, and injury. Rotate boars daily between breeding pens to reduce the influence of inactive or sterile boars.

Provide separate housing for new boars and isolate them for at least 60 days before exposure to the sow herd. This helps you detect disease and prevent herd infection.

Provide at least one boar for each four sows weaned/bred per week, which assumes that each sow is bred twice. Allow one or two additional boars for injuries or sickness. Another way to compute the number needed is to use a mature boar no more than twice a day and no more than ten times a week (use young boars no more than once a day and six times a week).

Provide additional boars if sows are weaned and bred as a group. Refer to Table 1.

COMBINING BUILDINGS

When combining buildings into a swine production system, consider the movement of three major products—feed, pigs, and manure. Animal movement includes:

- Sows from gestation-breeding to farrowing and back again.
- Pigs from farrowing to nursery, growing, and finishing.
- Gilts from growing-finishing to gestation-breeding.
- Market or feeder pigs and culled sows to market.

Place gestation-breeding on one side of farrowing and pig growing on the other side. Sows move easily between gestation-breeding and farrowing; pigs move easily from farrowing to the nursery, growing, and finishing buildings.

Several building arrangements are shown in Fig 24-28. Separate buildings work well for small producers; in-line or cross-farrow arrangements for small to medium producers; and 'H' or parallel arrangements for large producers. Most of these layouts are expanded by duplicating the entire building system.

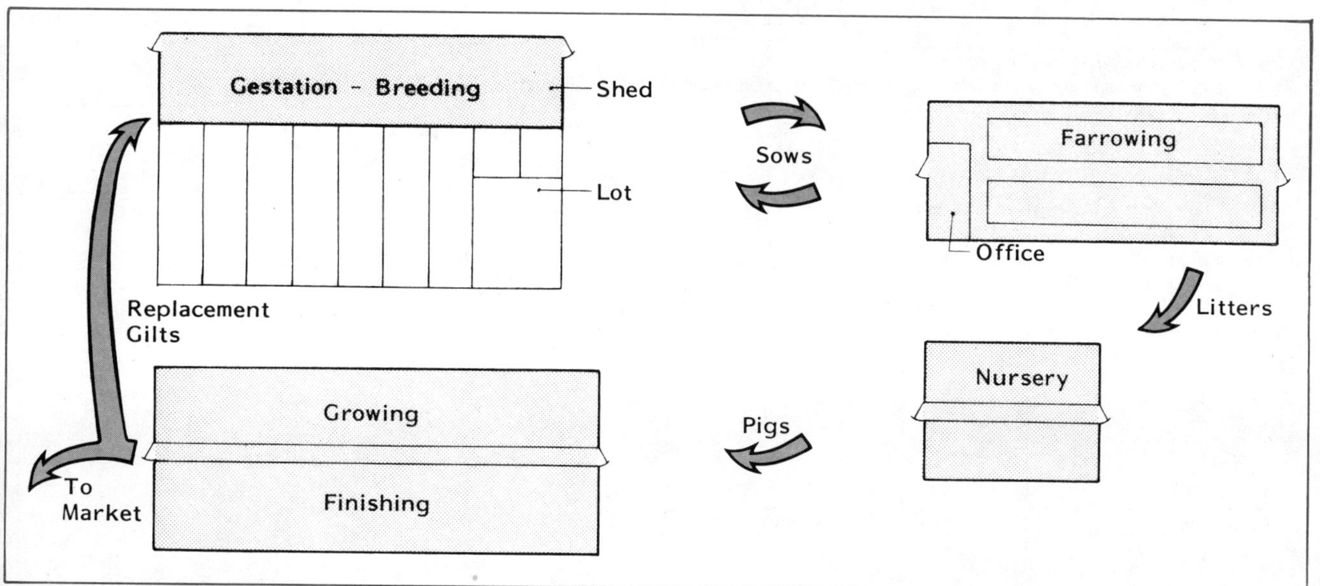

Fig 24. Separate buildings.
Moving animals from building to building is difficult. Expand by adding onto building lengths or duplicating entire building system.

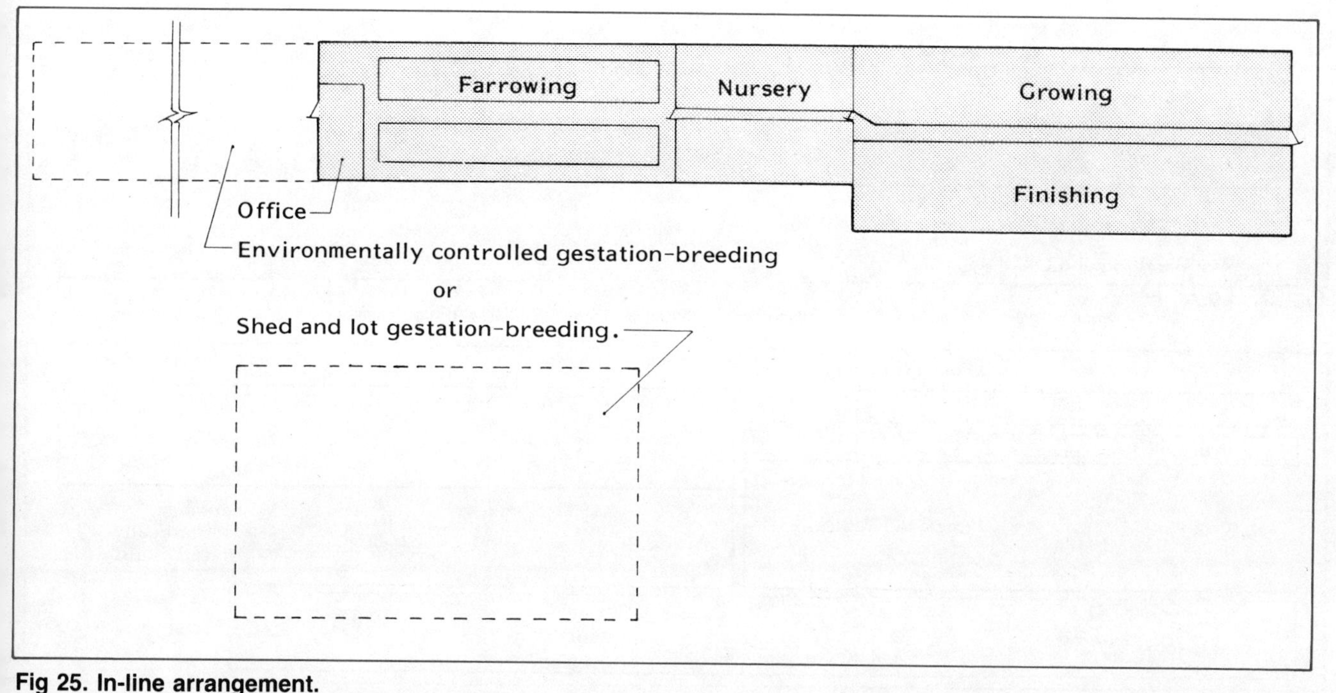

Fig 25. In-line arrangement.
Buildings are end to end so animals can be easily moved within the building. Gestation-breeding may be a separate facility or attached to the farrowing building.

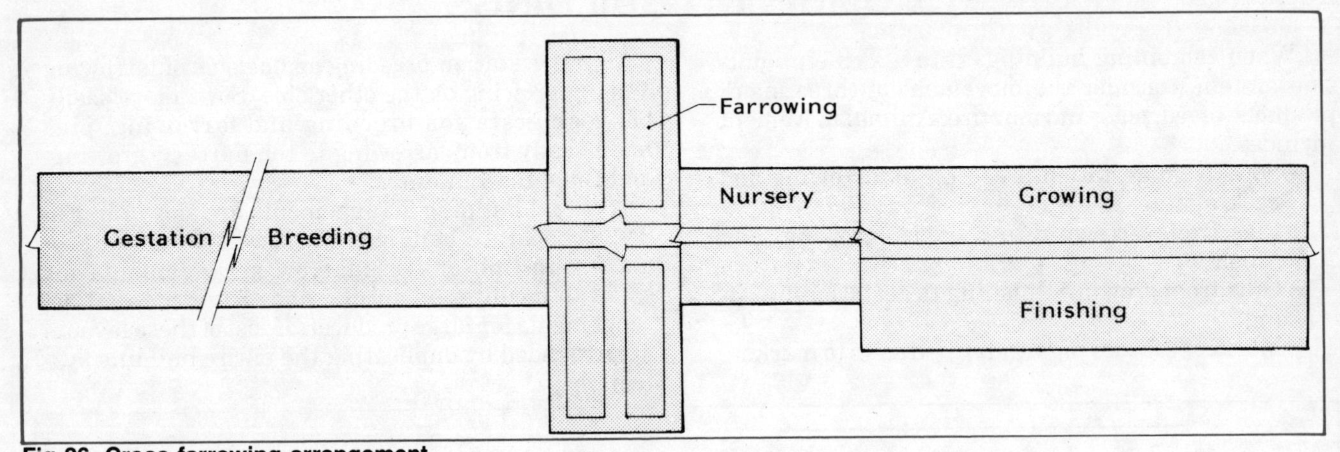

Fig 26. Cross-farrowing arrangement.
Similar to the in-line arrangement except farrowing is turned crossways to reduce overall building length.

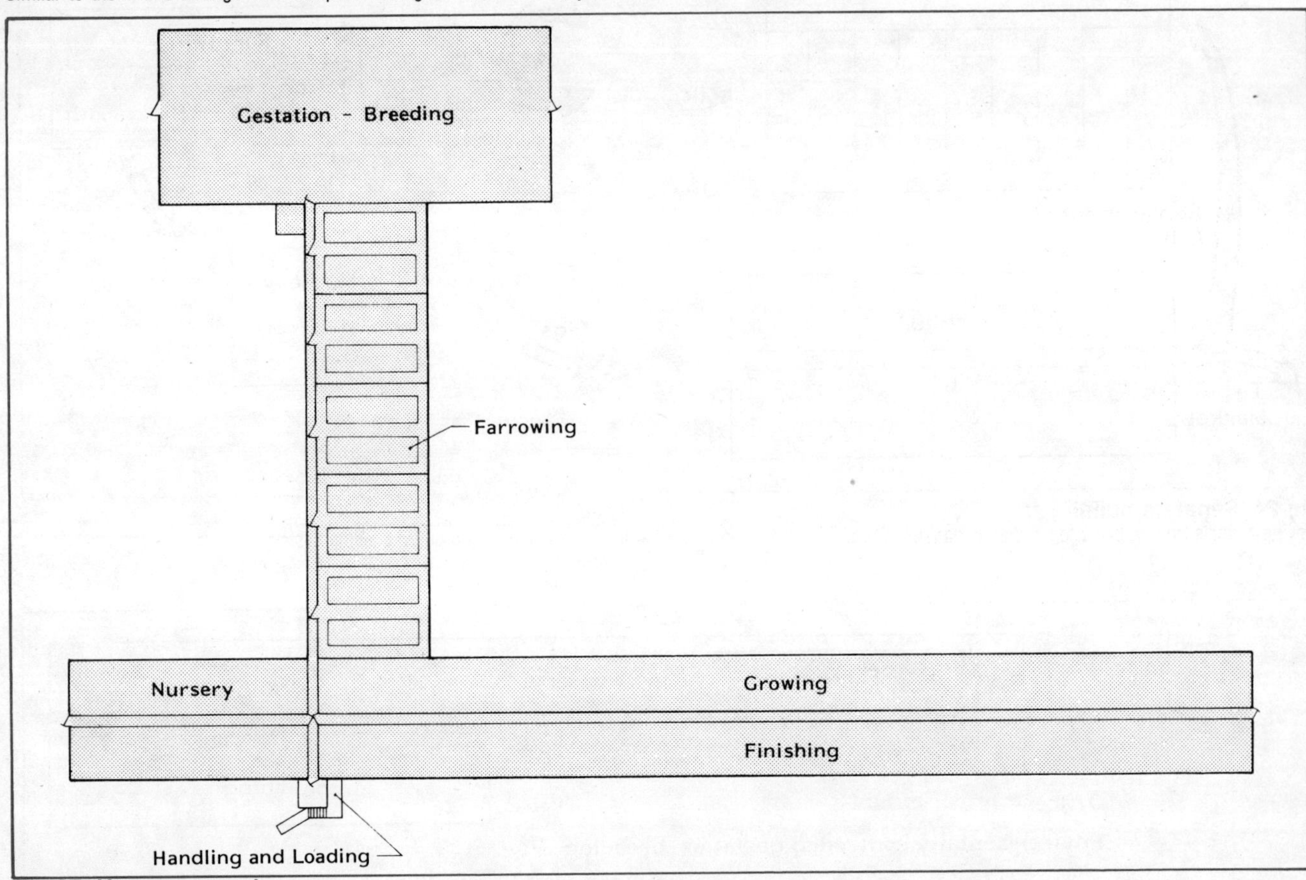

Fig 27. H arrangement.
Overall building complex length is reduced by placing farrowing lengthwise between gestation-breeding and pig growing.

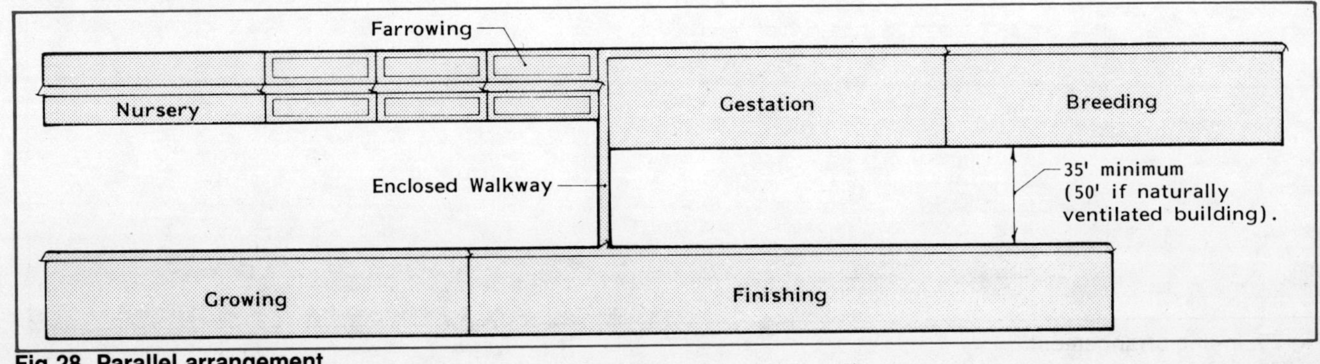

Fig 28. Parallel arrangement.
Parallel buildings are connected with enclosed walkways for easier animal movement.

Separate all buildings by at least 35′ for access, snow storage, and fire protection. Naturally ventilated buildings require 50′ of clearance. Leave room for future expansion of the building complex, utilities, and manure handling facilities.

Install fire walls and attic fire stops to restrict fires to one portion of the building instead of letting it spread throughout the entire complex, Fig 29. Use concrete block, mineral asbestos board, ¼″ magnesium oxychloride, or fire rated plywood for fire walls; mineral asbestos board, gypsum board, or metal for attic fire stops. Use fireproof materials for the inside and outside surfaces of enclosed walkways between buildings.

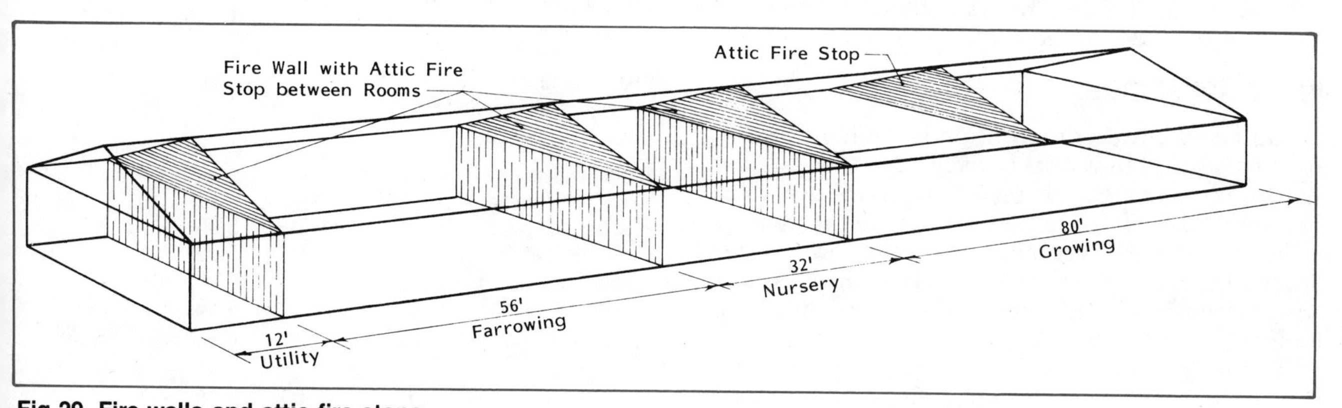

Fig 29. Fire walls and attic fire stops.
Construct fire walls between each building section. Construct attic fire stops above each fire wall but no more than 75′ apart.

PASTURE PRODUCTION

Producing swine on pasture requires relatively low investment in buildings and equipment. Year-round production is difficult in cold climates.

Recommendations for pasture production:

- Provide shade and shelter. Move portable houses periodically.
- Locate feeders at a well-drained site, preferably on pavement or wooden platforms.
- Arrange pasture lots along a permanent road. Prevent wallows near waterers by forming a curb around the slab. Slope the slab and surrounding area for quick drainage.

Pasture space requirements (depending on rainfall and soil fertility):

- 10 gestating sows/acre.
- 7 sows and litters/acre.
- 50 to 100 growing-finishing pigs/acre.

Shade space requirements:

- 15-20 ft²/sow.
- 20-30 ft²/sow and litter.
- 4 ft²/growing-finishing pig (to 100 lb).
- 6 ft²/growing-finishing pig (over 100 lb).

INSULATION

Insulation is any material that reduces heat transfer from one area to another. Some building materials, like wood, are good insulators, while others, like concrete and metal, are poor insulators.

Although all building materials have some insulation value, the term "insulation" usually refers to materials with a relatively high resistance to heat flow. The resistance of a material to heat flow is indicated by its **R-value,** with good insulators having high R-values. See Table 4.

Insulation Types

Batt and blanket insulation is available in 1″-8″ thicknesses and in widths to fit 16″, 24″, and 48″ stud spaces. Batts are 4′-8′ long, and blankets are up to 100′ long. Materials are fiberglass, mineral wool, or cellulose fibers. The batt or blanket may have a paper or aluminum face to serve as a partial vapor barrier. An additional vapor barrier is required in swine buildings.

Loose-fill insulation is packaged in bags and may be mineral wool, cellulose fiber, vermiculite, granulated cork, and/or polystyrene. It is easy to pour or blow above ceilings, in walls, and in concrete block cores. Poor insulation can settle in walls, leaving the top inadequately insulated.

Rigid insulation is made from cellulose fiber, fiberglass, polystyrene, polyurethane, or foam glass and is available in ½″-2″ thick by 4′ wide panels. Some types have aluminum foil or other vapor barriers attached to one or both faces. Rigid insulation can be used for roofs and walls or as a ceiling liner. It can also be used along foundations (perimeter insulation) or buried under concrete floors (if waterproof and protected from physical and rodent damage).

Support rigid insulation at least 2′ o.c. Use tongue-and-grooved panels or seal the joints with caulk or tape to prevent moisture from passing through the joints. Check for flammability and toxic gas production if burned. Check if your insurance company requires rigid insulation to be protected with fire-resistant materials.

Foamed-in-place insulation is usually obtained only through commercial applicators because it requires special equipment and experienced workers. Substandard application can cause excessive shrinkage. Apply a separate vapor barrier.

Sprayed-on insulation, applied to inside or outside surfaces, is difficult to protect with an adequate vapor barrier. Exterior application must also be protected from sunlight. As a result, improperly installed insulation may peel off.

Selecting Insulation

Consider the following factors:

- **Ease of installation.** Are you planning to do it yourself? Can it be done without tearing off the siding or inner wall surface? Some materials are harder to handle or take more time to install, which increases labor. Others are irritating to eyes and skin, requiring protective clothing and masks.

Table 4. Insulation values.
From 1981 ASHRAE Handbook of Fundamentals. Values do not include surface conditions unless noted otherwise. All values are approximate.

Material	R-value Per inch (approximate)	R-value For thickness listed
Batt and blanket insulation		
Glass or mineral wool, fiberglass	3.00-3.80*	
Fill-type insulation		
Cellulose	3.13-3.70	
Glass or mineral wool	2.50-3.00	
Vermiculite	2.20	
Shavings or sawdust	2.22	
Hay or straw, 20″		30+
Rigid insulation		
Exp. polystyrene,		
extruded, plain	5.00	
molded beads, 1 pcf	5.00	
molded beads, over 1 pcf	4.20	
Expanded rubber	4.55	
Expanded polyurethane, aged	6.25	
Glass fiber	4.00	
Wood or cane fiberboard	2.50	
Polyisocyanurate	7.04	
Foamed-in-place insulation		
Polyurethane	6.00	
Urea formaldehyde	4.00	
Building materials		
Concrete, solid	0.08	
Concrete block, 3 hole, 8″		1.11
lightweight aggregate, 8″		2.00
lightweight, cores insulated		5.03
Metal siding	0.00	
hollow-backed		0.61
insulated-backed, ⅜″		1.82
Lumber, fir and pine	1.25	
Plywood, ⅜″	1.25	0.47
Plywood, ½″	1.25	0.62
Particleboard, medium density	1.06	
Hardboard, tempered, ¼″	1.00	0.25
Insulating sheathing, 25/32″		2.06
Gypsum or plasterboard, ½″		0.45
Wood siding, lapped, ½″x8″		0.81
Windows (includes surface conditions)		
Single glazed		0.91
with storm windows		2.00
Insulating glass, ¼″ air space		
double pane		1.69
triple pane		2.56
Doors (exterior, includes surface conditions)		
Wood, solid core, 1 ¾″		3.03
Metal, urethane core, 1 ¾″		2.50
Metal, polystyrene core, 1 ¾″		2.13
Floor perimeter (per ft. of exterior wall length)		
Concrete, no perimeter insulation		1.23
with 2″x24″ perimeter insulation		2.22
Air space (¾″ to 4″)		0.90
Surface conditions		
Inside surface		0.68
Outside surface		0.17

*The R-value of fiberglass varies with batt thickness. Check package label.

- **What is to be insulated?** Will you be insulating a ceiling where many inches of material may be needed, or a wall or roof with thickness limitations?
- **Fire resistance.** Will the material require a fire-resistant liner to prevent rapid flame spread? Check with your insurance company before choosing insulation.
- **Animal contact.** Will the insulation be exposed to physical damage, requiring a protective covering which increases the cost?
- **Cost.** What will the different types of insulation cost considering preparation, installation, protection, etc., as well as purchase price? Variation in material and labor costs can be significant.

Insulation Levels

Minimum insulation values for swine buildings are shown in Fig 30 and Table 5.

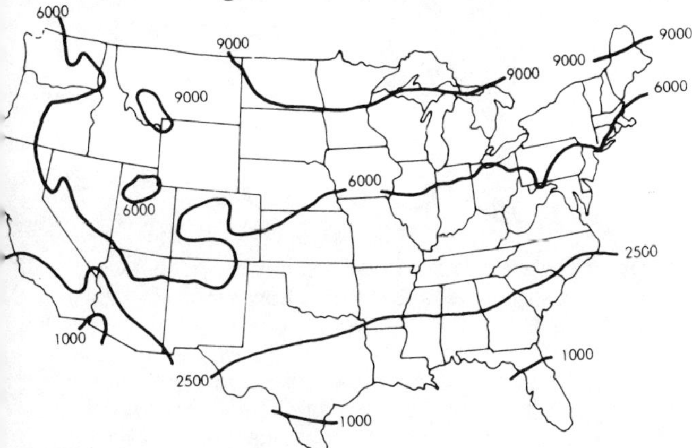

Fig 30. Winter degree days.
Accumulated difference between 65 F and average daily temperature for all days in the heating season.

Table 5. Minimum insulation levels for swine buildings.
R-values are for building sections.

	Recommended minimum R-values					
	Open front		Modified open front		Environment controlled	
Winter degree days	Walls	Roof	Walls	Roof	Walls	Ceiling
2500 or less	—	6	6	14	14	22
2501-6000	—	6	6	14	14	25
6001 or more	—	6	12	25	20	33

Installing Insulation

Fig 31-34 show common construction methods for insulated roofs, ceilings, walls, and foundations. Cracks around window and door frames, pipes, and wires result in cold spots, drafty areas, or condensation, and reduce ventilation inlet effectiveness. Caulk cracks and joints on outside surfaces.

Perimeter insulation reduces heat loss through the foundation and eliminates cold, wet floors. Insulate concrete foundations by covering the foundation exterior below the siding to a minimum of 16" below ground line. Use 2" rigid insulation and protect it with an impact-, moisture-, and rodent-resistant covering above ground.

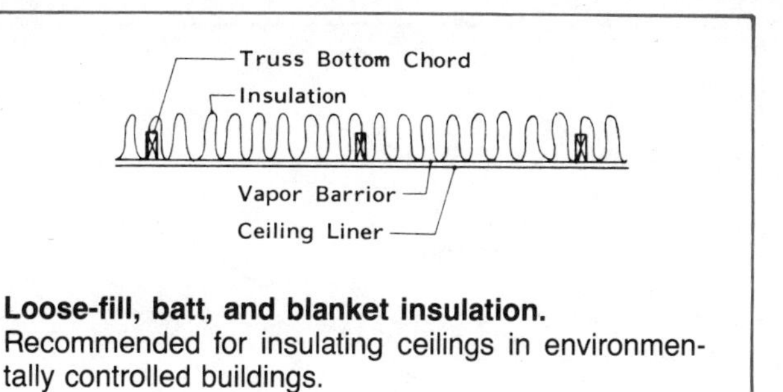

Loose-fill, batt, and blanket insulation. Recommended for insulating ceilings in environmentally controlled buildings.

Fig 31. Insulating ceilings.

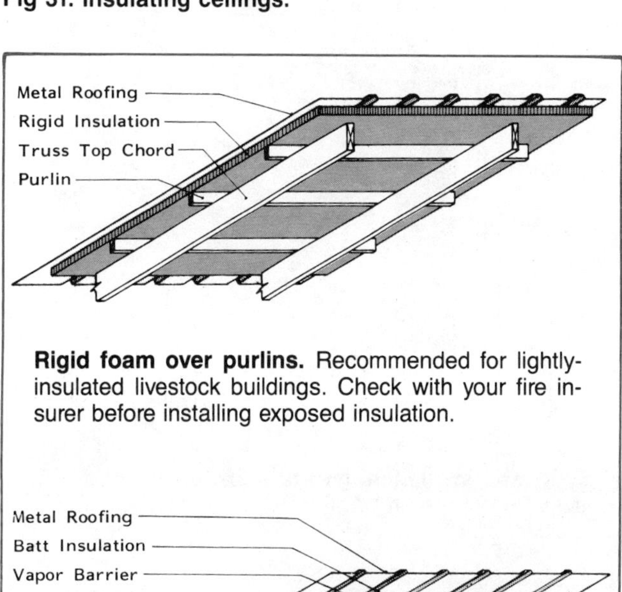

Rigid foam over purlins. Recommended for lightly-insulated livestock buildings. Check with your fire insurer before installing exposed insulation.

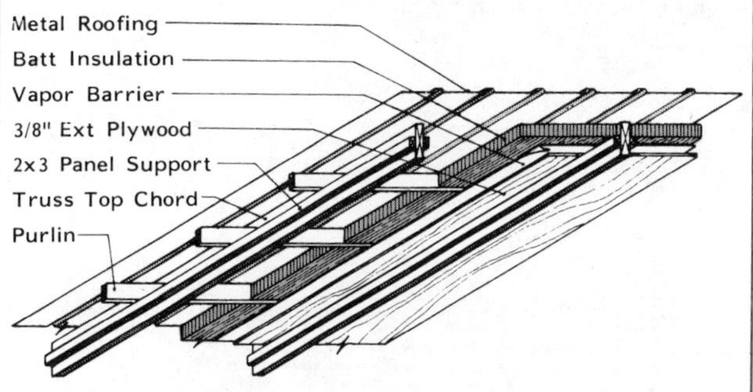

Insulated roof panels between trusses. Works well in modified open-front buildings. Fabricate on the ground and place between trusses. Apply roofing only after panels are nailed in place.

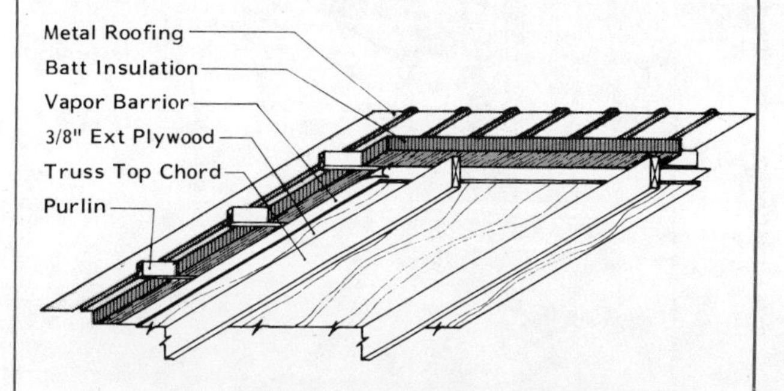

Insulated roof panels over trusses. Common in modified open-front buildings. Fabricate on the ground and lift into place—then apply roofing.

Fig 32. Insulating roofs.

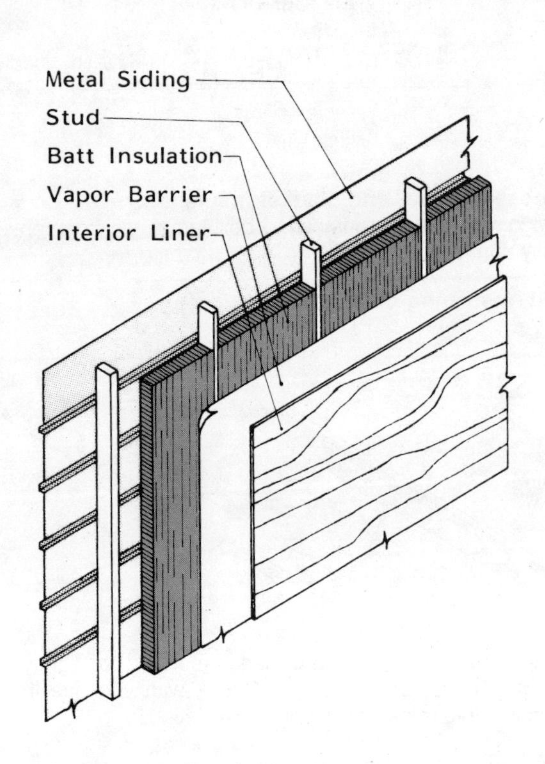

Stud wall insulation. (R = 12 if 2x4 studs, 20 if 2x6 studs). Common in environmentally controlled swine buildings.

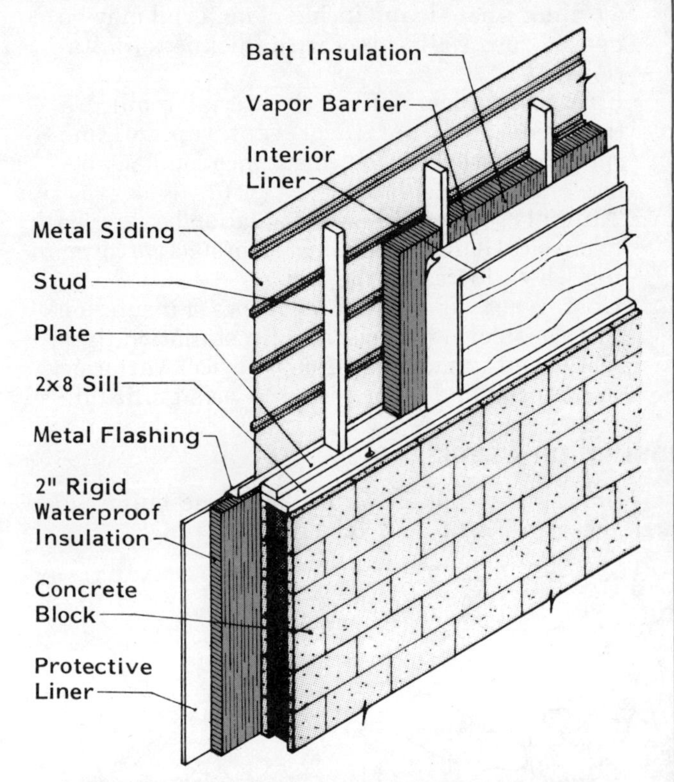

Concrete block and stud wall insulation. (Upper wall R = 12 if 2x4 studs, 20 if 2x6 studs. Lower wall R = 12 if standard blocks, 16 if lightweight blocks with cores filled). Concrete block wall (32" high) with stud wall above is popular for large swine. Although blocks provide a "pig-proof" wall, they lose much heat and "sweat" unless insulated. Providing adequate insulation levels is difficult with this wall.

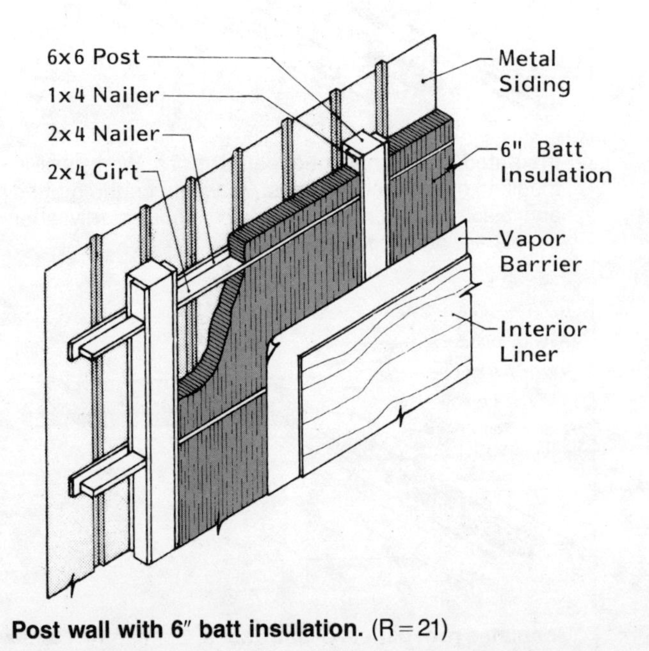

Post wall with 6" batt insulation. (R = 21)

2x2 Furring Board
Concrete Block
1½" Rigid Waterproof Insulation
Vapor Barrier
Interior Liner

Concrete block wall insulation. (R = 10 if standard blocks, 14 if lightweight blocks with cores filled).

Fig 33. Insulating walls.

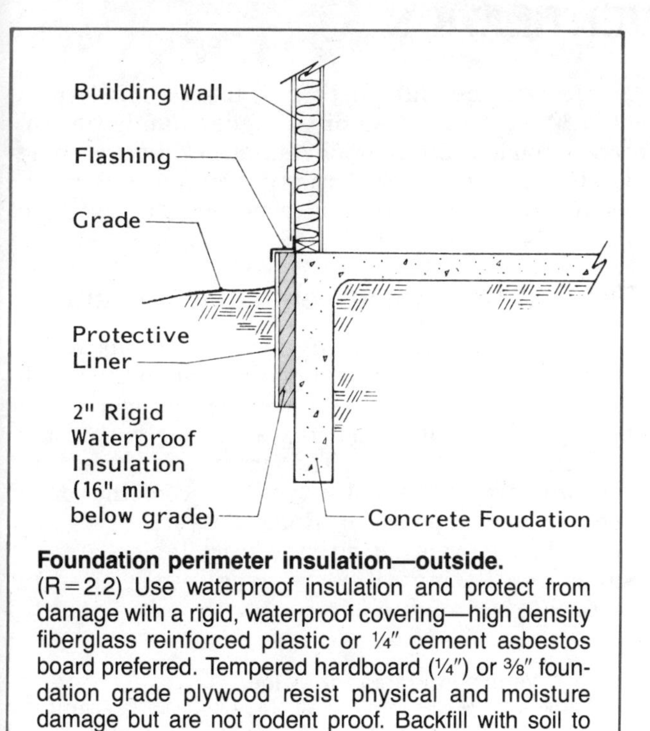

Foundation perimeter insulation—outside.
(R = 2.2) Use waterproof insulation and protect from damage with a rigid, waterproof covering—high density fiberglass reinforced plastic or ¼" cement asbestos board preferred. Tempered hardboard (¼") or ⅜" foundation grade plywood resist physical and moisture damage but are not rodent proof. Backfill with soil to within 6"-8" of insulation top.

Fig 34. Insulating foundations.

Table 6. Vapor barriers.
Based primarily on the 1981 ASHRAE Handbook of Fundamentals. A vapor barrier should have a perm rating less than 1.0.

Material	Perms*
Vapor barriers	
Aluminum foil, 1-mil	0.0
Polyethylene plastic film, 6-mil	0.06
Kraft and asphalt laminated building paper	0.3
Two coats of aluminum paint (in varnish) on wood	0.3-0.5
Three coats exterior lead-oil base on wood	0.3-1.0
Three coats latex	5.5-11.0
Common building materials	
Expanded polyurethane, 1"	0.4-1.6
Extruded expanded polystyrene, 1"	0.6
Tar felt building paper, 15-lb.	4.0
Structural insulating board, uncoated, ½"	50.0-90.0
Exterior plywood, ¼"	0.7
Interior plywood, ¼"	1.9
Tempered hardboard, ⅛"	5.0
Brick masonry, 4"	0.8
Poured concrete wall, 4"	0.8
Glazed tile masonry, 4"	0.12
Concrete block, 8"	2.4
Metal roofing	0.0

*1 perm = 1 grain of water/hr/ft^2/in. of mercury pressure difference.

Maximizing Insulation Effectiveness

Moisture Protection

Wet insulation increases heat loss and building deterioration. Install vapor barriers on the warm side of all insulated walls, ceilings, and roofs. Use vapor barriers underneath concrete floors and foundations to control soil moisture. Use waterproof rigid insulation if in contact with soil. Refer to Fig 31-34 for proper vapor barrier locations.

The ability of a material to allow water vapor to pass through it is called **permeability** and is measured in **perms.** A comparison of the permeability of various building materials is shown in Table 6. Select vapor barriers with a rating of less than 1.0 perm.

Doors and Windows

Use insulated and weatherstripped entry doors and locate them on the downwind side. For upwind entries, a 4' air lock between outer and inner doors prevents cold wind from blowing directly into the building when you enter or leave. To conserve floor area, consider making the air lock entrance a part of an office, storage area, or wash room.

Windows or skylights provide little or no benefit in heated livestock buildings. The insulation values of windows (R = 1-2.5) are well below an insulated wall (R = 13-15).

If remodeling a building, consider replacing all windows with insulated removable panels or permanent wall sections. Hinged or removable panels may be opened for summer ventilation.

Birds and Rodents

To prevent rodent damage, cover exposed perimeter insulation with a protective liner and maintain a rodent bait program.

To prevent bird damage, construct buildings so birds can't roost near the insulation and consider screening all vent openings. Use ½"x½" hardware cloth for air intake vent openings and ¾"x¾" hardware cloth for air outlet vent openings. Screened vent **outlets** are very susceptible to freezing shut in cold climates and you may need to knock ice off the screen regularly during cold spells. An aluminum foil covering is not sufficient protection, but may make the insulation less attractive to birds.

Dead pigs left where wild animals can feed on the carcasses can be a major disease transmission source.

Fire Protection

The rate at which fire moves through a room depends on the interior lining material. Many plastic foam insulations have high flame spread rates. To reduce risk with these materials, protect them with fire-resistant coatings. Materials which provide satisfactory protection include:

- ½" thick cement plaster.
- Fire rated gypsum board (sheet rock). Don't use in high moisture environments such as animal housing.
- ¼" thick sprayed-on magnesium oxychloride (60 lb/ft^3) or ½" of the lighter, foam material.
- Mineral asbestos board ⅛"-⅜" thick.
- Fire rated ½" thick exterior plywood.

MECHANICAL VENTILATION

A mechanical ventilating system has fans, controls, and air inlets or outlets. It provides more control over room temperature and air movement than natural ventilation. Mechanical ventilation works especially well in farrowing and nursery rooms, where young animals are susceptible to low temperatures, sudden temperature changes, and drafts.

Principles of Ventilation Design

Temperature Control

Ventilation removes moisture and odor in cold weather and excess heat the rest of the year. To maintain a constant room temperature, the heat **produced** by animals and heaters has to equal heat **lost** through building surfaces and ventilation.

If heat loss exceeds animal heat production, provide supplemental heat. If heat production exceeds heat loss, increase ventilation.

Moisture Control

During cold weather, ventilation brings cold, relatively dry air into the building. The air is warmed by energy from animals, electrical equipment, and supplemental heat. As the air temperature rises, it can hold more moisture and its relative humidity decreases. The moisture holding capacity of air nearly doubles for every 20 F rise in temperature. This ventilating air picks up moisture from the building (which increases its relative humidity) and exhausts it from the building at a high relative humidity. Fig 35.

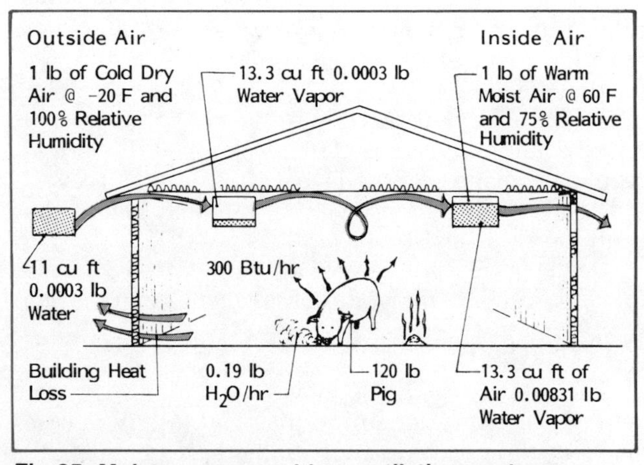

Fig 35. Moisture removed by ventilating systems.

The amount of moisture to be removed depends on animal type, animal size, and manure handling system. Generally, larger animals give off more moisture and require more moisture control ventilation. With totally slotted floors, most liquids will drain through instead of evaporating, reducing the moisture load on the ventilating system by about 50%, compared with solid floors. Partly slotted floors remove about 15% of the moisture, compared with solid floors.

Design ventilation to maintain room air at 50%-80% relative humidity. Higher humidities increase condensation; lower humidities increase dust levels. Also, a 50%-80% relative humidity is detrimental to the airborne bacteria found in livestock buildings.

Types of Mechanical Ventilating Systems

An effective ventilating system automatically moves the required amount of air through the building to control temperature, moisture, and odor; and it properly distributes and mixes the air within the building without drafts. The system may be a positive pressure type, exhaust type, or a combination of the two (neutral pressure). Regardless of the type, the recommended ventilating rates from Table 7 are the same. Ventilating rates are usually measured in cubic feet per minute (cfm).

Table 7. Ventilation rates for swine.
One-half of the hot weather ventilation rate can be accomplished with circulation fans rather than exhaust fans.

Animal type	Weight lb	Cold weather rate	Mild weather rate*	Hot weather rate*
		- - - - - - - - cfm/hd	- - - - - - - -	
			(additional = total)	
Sow and litter	400	20	+60 = 80	+420 = 500
Prenursery pig	12-30	2	+ 8 = 10	+ 15 = 25
Nursery pig	30-75	3	+12 = 15	+ 20 = 35
Growing pig	75-150	7	+17 = 24	+ 51 = 75
Finishing pig	150-220	10	+25 = 35	+ 85 = 120
Gestating sow	325	12	+28 = 40	+110 = 150**
Boar	400	14	+36 = 50	+250 = 300

* Add the value after the "+" sign to the previous column to get the total mild or hot weather rate. For example, a sow and litter require 20 cfm in cold weather, (20 + 60) = 80 cfm in mild weather, and (80 + 420) = 500 cfm in hot weather.

** +260 = 300 cfm as the hot weather rate for gestating sows in a breeding facility.

Positive Pressure Ventilating Systems

Fans force fresh air into the building, where it is distributed with a baffle in front of the fan or a duct having carefully placed outlets, Fig 36. Use Table 10 to size the duct. Insulate the duct to prevent condensation.

Size positive pressure exhaust outlets at 300 in^2 / 1000 cfm of fan capacity. For example, provide one pair of 2" diameter holes for each 20 cfm or one pair of 3" diameter holes for each 45 cfm of airflow through the duct. Base number of holes on the highest airflow through the duct, and block off part of each hole when operating at lower airflow rates. Space holes uniformly along the duct.

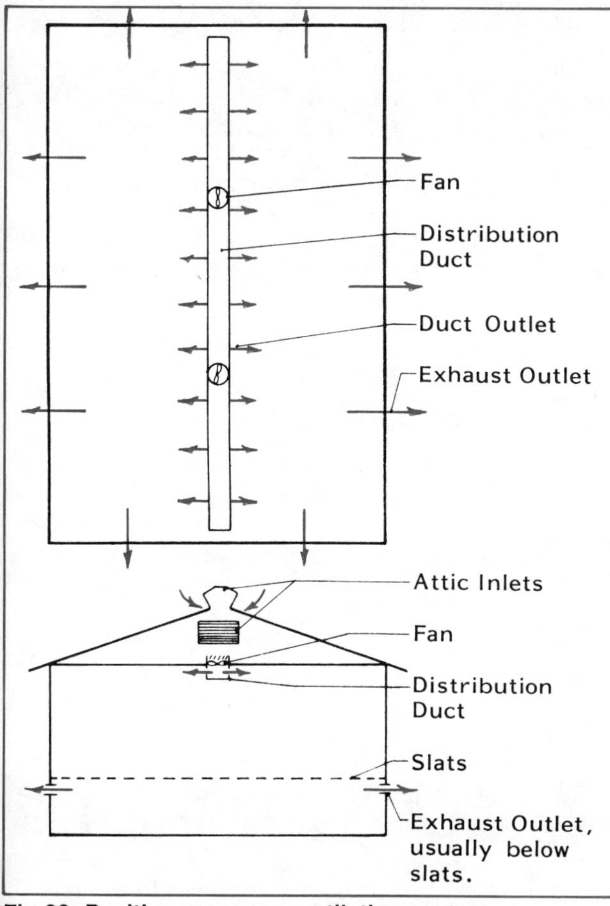

Fig 36. Positive pressure ventilating system.

Positive pressure systems tend to force moist air into the walls and attic, so a good vapor barrier is important. Also, frost forms at air leaks around doors, freezing them shut. Positive pressure systems are often installed in remodeled buildings, because pressure outlets are easier to construct than inlets for exhaust ventilation.

Exhaust Ventilating Systems

Exhaust fans force air out of the building, creating a negative pressure, which draws fresh air into the building through baffled inlets, Fig 37. If the inlets are sized and located properly, fresh air is distributed uniformly throughout the room. Exhaust is by far the most common ventilating system in swine housing.

Neutral Pressure Systems

Neutral pressure systems include inlet shutters, an air distribution duct (or tube), a duct fan, and exhaust fans, Fig 38. In cold and mild weather, ventilating air is distributed through the duct and exhausted through wall fans. A motorized shutter, located in the wall near the duct fan, allows fresh air to enter the duct from the outside or from the attic, Fig 39. The duct fan usually moves more air than is required in cold weather, so the shutter opens intermittently to let in the required amount of air. An exhaust fan operates when the shutter opens, to draw fresh air in through the shutter. When the intake

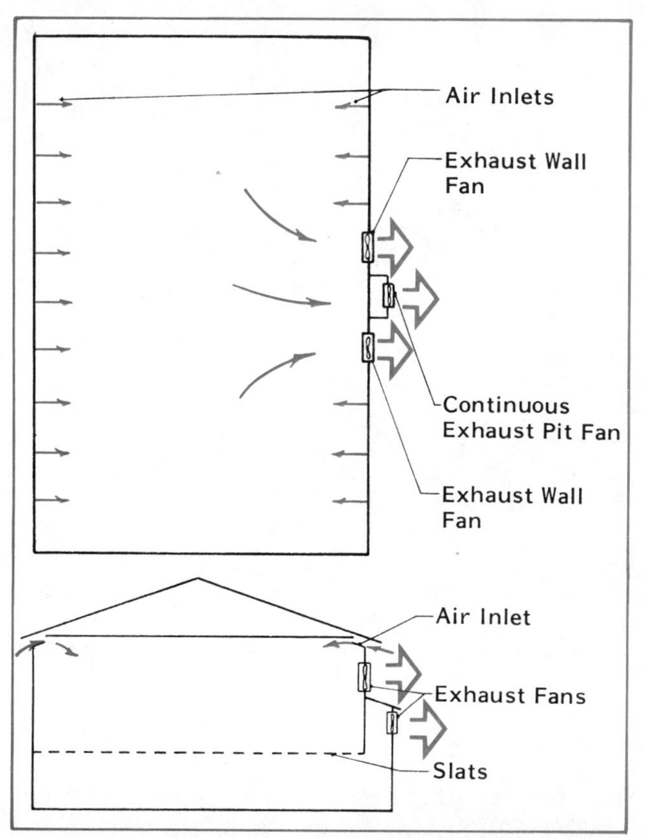

Fig 37. Exhaust (negative pressure) ventilating system.

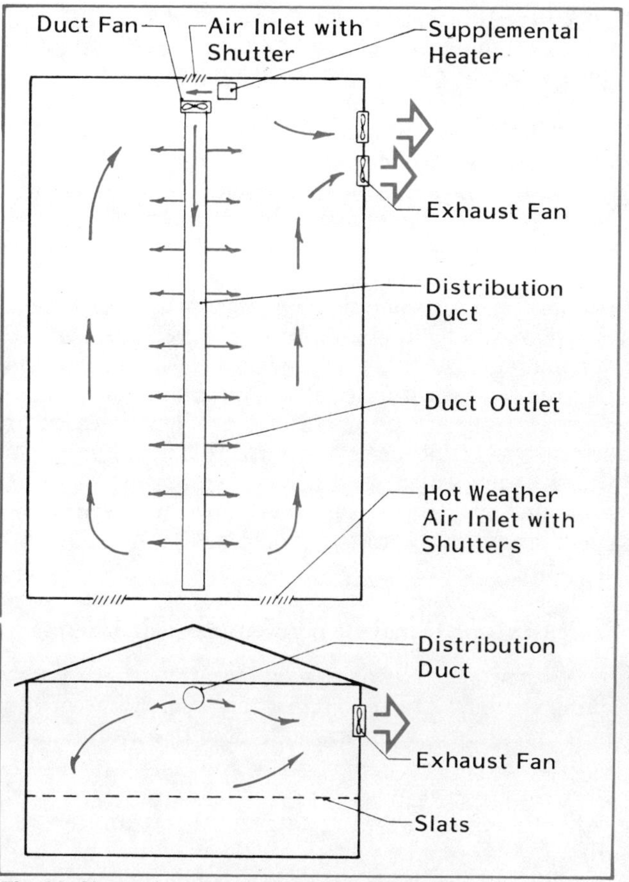

Fig 38. Neutral pressure ventilating system.

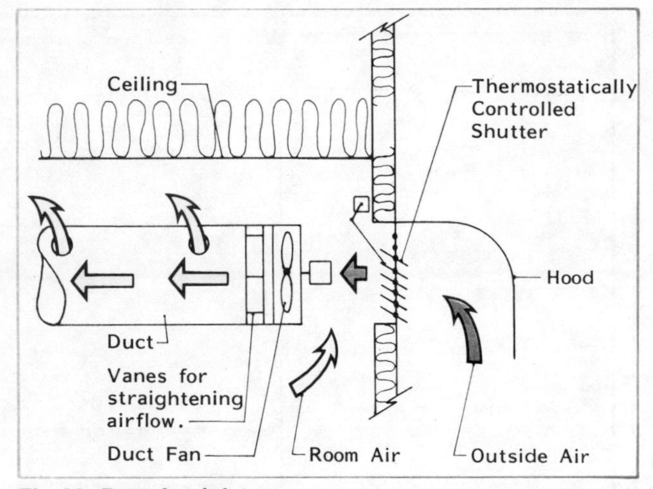

Fig 39. Duct fan inlet.
Remove the top one or two leaves from the shutter in case of power or freezing failure.

shutter closes, the duct fan circulates only room air through the distribution duct. If unit space heaters are required, locate them at the duct intake so the duct will distribute the heat.

Design duct and duct fan to handle at most the mild weather ventilating rate. Additional inlets and exhaust fans are required to supply the hot weather rate. Locate the hot weather inlets (usually motorized shutters) at the end of the building opposite the exhaust fans. Locate young animals so incoming air will not chill them.

While this system generally provides good air distribution at low ventilating rates, there are problems that can result from recirculating air:

- Dust collects in the duct.
- The intake shutters may freeze.
- Increased air movement can create a draft.
- Use of the same air duct from one animal group to another may transfer dust and bacteria.

The only solution to dust buildup is periodic cleaning of the duct. This may mean partial disassembly, so consider cleaning ease when constructing homemade ducts. Shutters may freeze if no supplemental heat is provided or if the duct fan is too far from the shutters to move warm air past them. Shutter freeze-up is a serious problem because it restricts ventilation. If you have this problem, keep the shutters warm with a heat lamp. Correct draft problems by installing a lower capacity or variable-speed duct fan, by raising room temperature, or by installing hovers.

Exhaust Ventilating Systems—Air Inlets

With proper inlet design and location, there are no damp corners, drafts, or dead air spots. One of the best air inlets is a continuous slot with an adjustable baffle. The baffle restricts the inlet opening to increase air velocity and improve air mixing. Use rigid baffles to avoid warping and resulting uneven air distribution. Tightly seal all doors, windows, and unplanned openings, especially when operating at low ventilating rates.

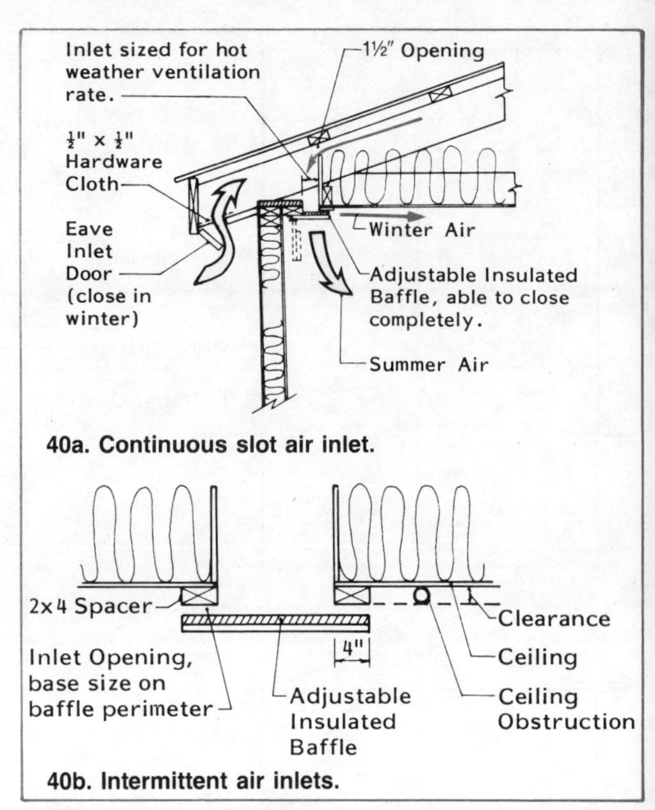

40a. Continuous slot air inlet.

40b. Intermittent air inlets.

Fig 40. Baffled air inlets for "across-the-ceiling" airflow.
Do not mount pipes or lights within 4′ of inlets—a smooth ceiling prevents cold air from being deflected down onto animals.

Intermittent air inlets are common in remodeled buildings because they are easier to build than continuous slots, Fig 40b. Space intermittent openings uniformly in the room, but place them at least 8′ from building walls and at least 16′ apart. Intermittent inlet size depends on required airflow. Construct the baffle to extend 4″ beyond all sides of the inlet opening.

While most exhaust ventilating systems with continuous slot inlets force air across the ceiling, Fig 40, a "down-the-wall" pattern also works. Fig 41. Air flows directly into the pig zone, so control drafts with hovers over pig sleeping areas and solid ends and creep dividers on farrowing stalls.

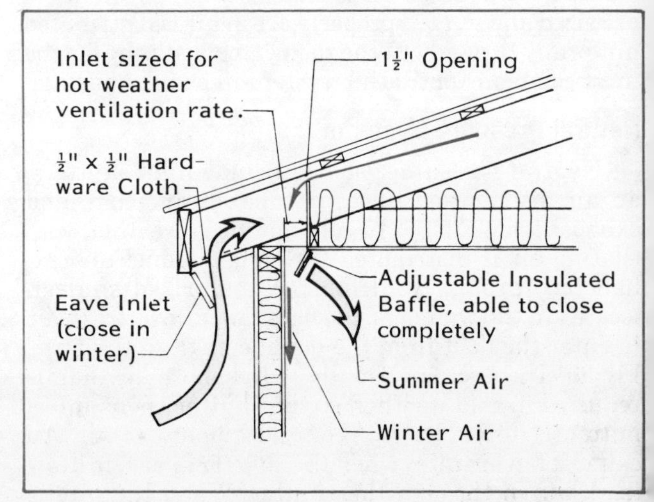

Fig 41. Baffled air inlet for "down-the-wall" airflow.

Inlet Location

For buildings up to 35′ wide, place slot inlets at the ceiling along both sidewalls; for wider buildings add one or more interior ceiling inlets, Fig 42.

Air should travel less than 75′ from inlet to fan. During cold weather, close and seal air inlets within 8′ of the winter exhaust fans.

Inlet Size

Size inlets to maintain a high velocity (700-1000 fpm) where air leaves the baffle. If the velocity is less, cold air settles too rapidly and may chill the animals; if greater, only the top portion of the room is ventilated and high static pressure decreases fan capacity. Usually, inlets are sized for the hot weather rate, then reduced with a baffle for lower ventilating rates. A 1″ wide, 1′ long slot delivers 50 cfm at 700-1000 fpm. Similarly, a 2″ wide, 1′ long slot delivers 100 cfm at 700-1000 fpm.

Example: Calculate proper slot openings for a 24′x84′, 30-sow farrowing building. Group exhaust fans in the center of the south wall and locate inlets at the ceiling on the long sides of the room.

1. Total slot inlet length is: (building length x 2):
   (84′ x 2) = 168′.
   In cold weather, close the inlets over the continuous exhaust fans to prevent short circuiting of air. **Winter** slot opening is about 150′.
2. Maximum ventilating rate per foot of inlet is: (hot weather rate, from Table 7 x no. of sows) ÷ total slot inlet length:
   (500 cfm x 30 sows) ÷ 168′ = 89.3 cfm/ft.
3. A 1″ wide slot provides 50 cfm per foot of length, so hot weather slot width is:
   (89.3) ÷ (50 cfm/in.) = 1.8″ or about 2″ wide.
4. In cold weather, the ventilating rate is much less: (cold weather rate, from Table 7 x no. of sows) ÷ total slot inlet length:
   (20 cfm x 30 sows) ÷ 150′ = 4.0 cfm/ft.
5. To provide this cold weather rate, a slot must be partially closed with the baffle, to:
   (4.0) ÷ (50 cfm/in.) = 0.08″.
   In practice, unplanned air leaks make it difficult to maintain adequate air velocity at low ventilating rates. Decrease the inlet opening by 30%-50% to account for these air leaks.

The cold weather rate may be so low that it is difficult to close the baffle to a small enough opening. So, close every other 4′ section of inlet and double the slot opening in the alternate 4′ sections.

Cold Weather Inlets

In winter, bring fresh air from the attic. Run moisture control fans **continuously,** because when fans are off, warm room air rises through the slot, condenses on the underside of the cold roof, and drips on attic insulation. Avoid timers if possible. Air enters the attic through downwind soffit openings, gable louvers, or ridge ventilators, Fig 43.

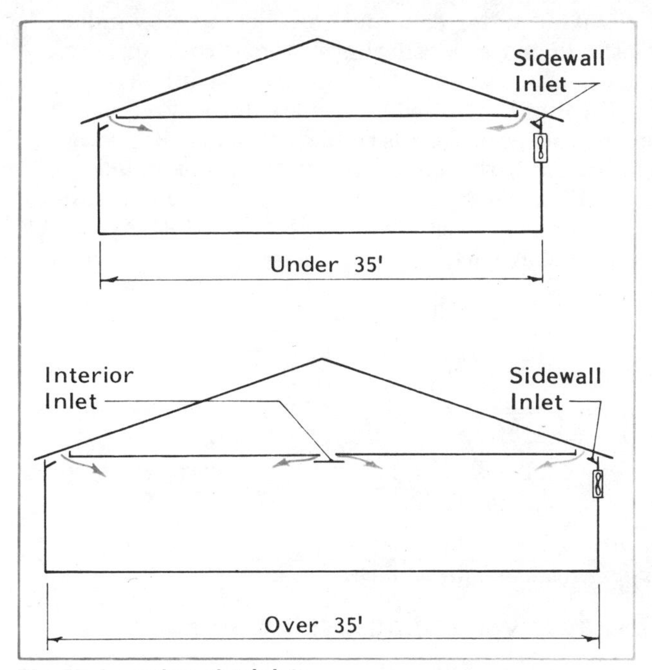

Fig 42. Locating slot inlets.

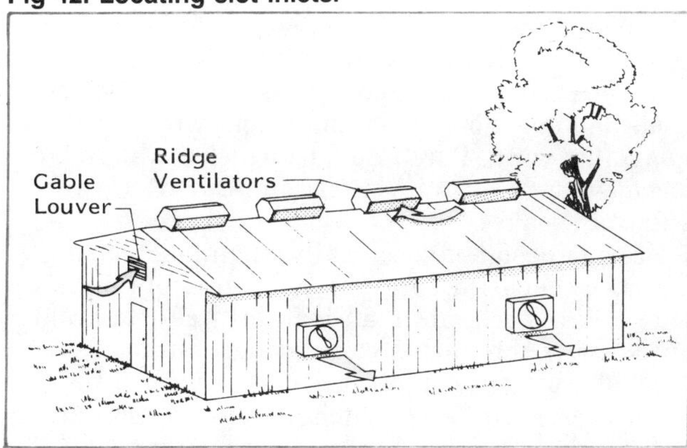

Fig 43. Gable louvers and ridge ventilators.
Screen attic openings with ½″x½″ hardware cloth. Total attic opening, ft² = (Mild weather ventilation, cfm) ÷ 200.

Hot Weather Inlets

In hot weather, pull fresh air directly from the outside instead of the attic. To do this, open the doors in both eaves, Fig 40a. Some air entering the eaves ventilates the attic to lower the ceiling temperature. Screen attic openings with ½″x½″ hardware cloth. Smaller mesh screen plugs rapidly with dust and restricts air flow.

Inlet Control

Air inlet control is critical to good ventilation. The size of the slot inlet should change each time the ventilating rate changes, preferably automatically. Operate manually-adjusted baffles from one location with a winch and cable system. Install a manometer next to the winch for more accurate baffle adjustments. A manometer measures static pressure, which is the difference in atmospheric pressure inside and outside a building. A slot width sized to deliver air at 700-1000 fpm creates a static pressure of

0.04"-0.06" water column. Keep doors and windows closed when the ventilating system is running, or air may not be drawn in through the slot inlets.

Curtain-controlled slot inlets are self-adjusting, but are not as precise as rigid baffles, Fig 44. Condensation and frost may collect on the uninsulated curtain. Plastic curtains tend to become less flexible with age, so check them at least once a year and replace as needed.

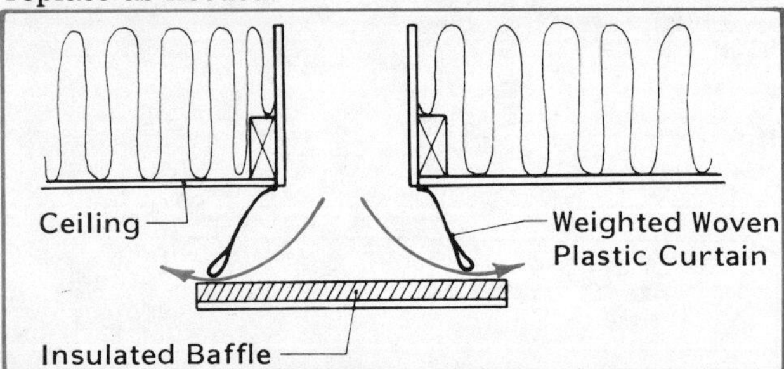

Fig 44. Gravity curtain inlet.

Exhaust Ventilating Systems—Fans

Fan Selection Guidelines

Select fans capable of moving the required amount of air against at least ⅛" static pressure. Variable-speed fans have poor pressure ratings at low speeds and may not deliver enough air when operating against wind. Purchase fans which have an Air Movement and Control Association (AMCA) "Certified Rating" seal.

Use fans designed specifically for animal housing. Buy totally enclosed, split-phase or capacitor type fan motors. Wire each fan to a separate circuit to avoid shutdown of the entire ventilating system if one motor blows a fuse. Protect each fan with a fused switch at the fan—size switch at 25% over fan amperage.

Fan Location

With a rigid baffle slot inlet, fan location has little effect on air distribution. If possible, place winter fans in the downwind wall. To prevent air from becoming too stale, place fans no more than 75′ from the farthest inlet.

If fans must exhaust against prevailing winter winds, protect them with a windbreak 5′-10′ from the building, Fig 45. Put anti-backdraft shutters on all non-continuous fans. Place shutters on the inside (animal side) of fans to decrease freezing problems. Space mechanically ventilated buildings at least 35′ apart, so fans don't blow foul air into the intakes of adjacent buildings.

Fan Controls

Choose thermostats designed especially for livestock housing with an off-on range of 5 F or less. Locate thermostat:

- At or near center of building width.
- Out of animal reach.
- Away from cold walls and ceilings.
- About 4′-5′ off the floor.
- Out of the path of furnace exhausts, inlet air, and direct sunlight.

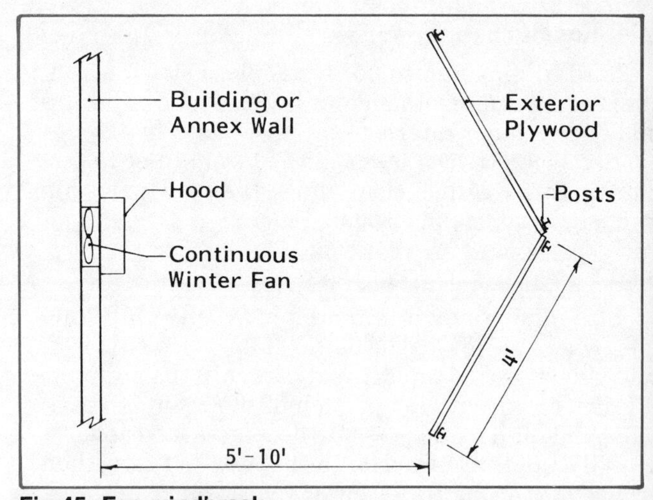

Fig 45. Fan windbreak.
Windbreaks for continous fans help maintain more uniform ventilation. They are especially useful for variable-speed fans operated at the bottom half of their capacity range.

Determining Fan Sizes and Thermostat Settings

Required ventilation varies with outside temperature. Select fan capacities and thermostat settings for a **gradual** airflow increase with an outside temperature increase. Use several fans, multi-speed fans, or variable-speed fans. Find the recommended ventilating rates in Table 7, then select fan sizes from product literature.

Most thermostats have an on-off range of about 3-5 F, and fans are normally staged to come on in 2-4 F increments.

Example: For the 30-sow farrowing building of the previous example, Table 7 gives the following ventilating rates:

- Cold weather: 30x20 = 600 cfm total
- Mild weather: 30x60 = 1800 cfm (Total = 1800 + 600 = 2400 cfm)
- Hot weather: 30x420 = 12600 cfm (Total = 12600 + 2400 = 15000 cfm)

Choose the desired room temperature from Table 8 (70 F). Set heater thermostats about 2 F below the desired room temperature (68 F). With a 5 F range on the furnace thermostat, the furnace shuts off at 73 F. The cold weather fan runs continuously so it does not have a thermostat. Set the thermostat of the smallest mild weather fan at about 3 F above the furnace-off temperature (76 F). Make sure heater does not operate after mild weather fans turn on. Set the other warm weather fans to turn on at increasing increments of 2-4 F (78 F, 80 F, etc.). The more fans in the room, the smaller the increments.

If variable-speed fans are used for cold weather ventilation, adjust the minimum speed to the cold weather rate (600 cfm). Set the fan controller to begin increasing speed at about 3 F above the furnace off temperature (76 F). Most variable-speed fans increase speed over a range of about 5 F. The cold weather variable-speed fan runs continuously so it does not have a thermostat. Except for totally slotted floors over storage pits, an emergency thermostat can be installed to shut the fan off at 35 F. Set the other fans to come on in 2-4 F increments, starting at about 5 F higher than the variable-speed fan setting (81 F, 83 F, etc.).

Heaters

Table 8 gives the supplemental heat requirements for typical livestock housing in the North Central U.S. Swine heating systems include radiant, floor, unit space, and air make-up heaters.

Table 8. Supplemental heat requirements.
Sized for a moderately well insulated building ventilated at twice the recommended cold weather rate. Additional creep heat is needed for young pigs in farrowing and nursery. Size creep heaters at about half the table value.

Animal type	Inside temp, F	Supplemental heat, Btu/hr/hd Slotted floor	Bedded/ scraped floor
Sow and litter	80	4000	—
	70	3000	—
	60	—	3500
Prenursery pig (12-30 lb)	85	350	—
Nursery pig (30-75 lb)	75	350	—
	65	—	450
Growing-finishing pig (75-220 lb)	60	600	—
Gestating sow/boar	60	1000	—

Radiant Heaters

Radiant heaters work well in creep and sleeping areas. They allow the pig to find comfort by moving closer to or farther from the heater. Energy from a radiant heater passes through air without warming it—upon striking the pig or other surface, the energy is absorbed and the pig or surface is warmed.

Infrared heaters produce radiant heat from gas or electricity. Many types are available, from heat lamps to sophisticated fan-driven pipe units which can heat an entire building.

With floor heat in the farrowing creeps, provide 250 watt (852 Btu/hr) overhead heat lamps for the first few days after farrowing. If no floor heat is used, provide 2200 Btu/hr of overhead radiant heat per litter.

Heat lamps are a potential fire hazard. Suspend them on chains and make the lamp cord 1′ shorter than the floor-to-ceiling height. Mount lamps at least 30″ above pen floors and 18″ above creep floors. Place no more than seven 250 watt heat lamps on one 20 amp circuit.

Catalytic radiant heaters are flameless and have relatively low surface temperatures. No electricity is required and they are relatively unaffected by dust and moisture. Catalytic heaters are usually not thermostatically controlled, which makes them less efficient.

They consume LP gas but require no special venting in a properly ventilated building. Never ventilate at a rate below 4 cfm/1000 Btu/hr of unvented heater capacity or a buildup of poisonous combustion gases may result.

Floor Heat

Floor heaters are electric resistance cables or hot water pipes buried in concrete, Figs 46 and 47, or electric heating coils in fiberglass pads placed over an existing floor. Floor heat evaporates liquids from the floor surface which increases relative humidity. Arrange resting areas, waterers, and dunging areas so the heated area is seldom wet. Do not heat the floor under the sow in farrowing buildings.

Heat the right amount of floor area, Table 9. Too much heated area wastes heat; too little encourages pigs to pile. Use waterproof insulation underneath a heated floor but do not use insulation or excessive bedding on heated sleeping areas. Do not lay the heating elements directly on plastic insulation. Install perimeter insulation if the heated slab is near an outside wall. Uniform concrete thickness above hot water pipes or electric heat cables prevents hot spots.

Table 9. Heated floor design criteria.

Pig weight lb	Heated floor space ft^2	Floor surface temperature F	Hot water pipe spacing in.	Electric heat W/ft^2
Birth-30	6-15/litter	85-95	See Fig 46	30-40
30-75	1-2/pig	70-85	5	25-30
75-150	2-3/pig	60-70	15	25-30
150-220	3-3½/pig	50-60	18	20-25

Locate the temperature sensing bulb about 1″ below the heated floor surface and about 4″-6″ from a heat pipe or 2″ from an electric heat cable. Install the temperature sensing bulb and connecting tube in a 1″ conduit or pipe so they can be removed for maintenance. Use a large radius bend in the conduit or pipe so the bulb can be inserted easily.

For farrowing buildings, provide about 95 F slab surface temperatures at farrowing. Lower gradually to 85 F when pigs are 3 weeks old. Heat lamps or hovers are common in addition to floor heat for 3 to 5 days after farrowing. Floor heat is occasionally installed in nursery buildings and naturally ventilated, growing-finishing buildings. See Table 9 for desired floor temperatures for larger pigs.

To calibrate the thermostat (which indicates internal instead of surface slab temperatures), adjust it to obtain desired surface temperatures, using a thermometer laid on the floor.

Start heating the floor at least 2 days before it is needed, to allow time for the concrete to warm up. Floor heat alone seldom provides the total heat requirements of a building.

Hot water floor heat uses wrought iron, black iron, or high temperature (160 F) plastic pipe. Do not use galvanized or cold water plastic pipe. Fig 46 shows the necessary equipment for a hot water floor heating system.

An input of 50 Btu/hour for each foot of pipe keeps the floor temperature at about 95 F. Multiply the linear feet of pipe times 50 Btu/hr to get the required

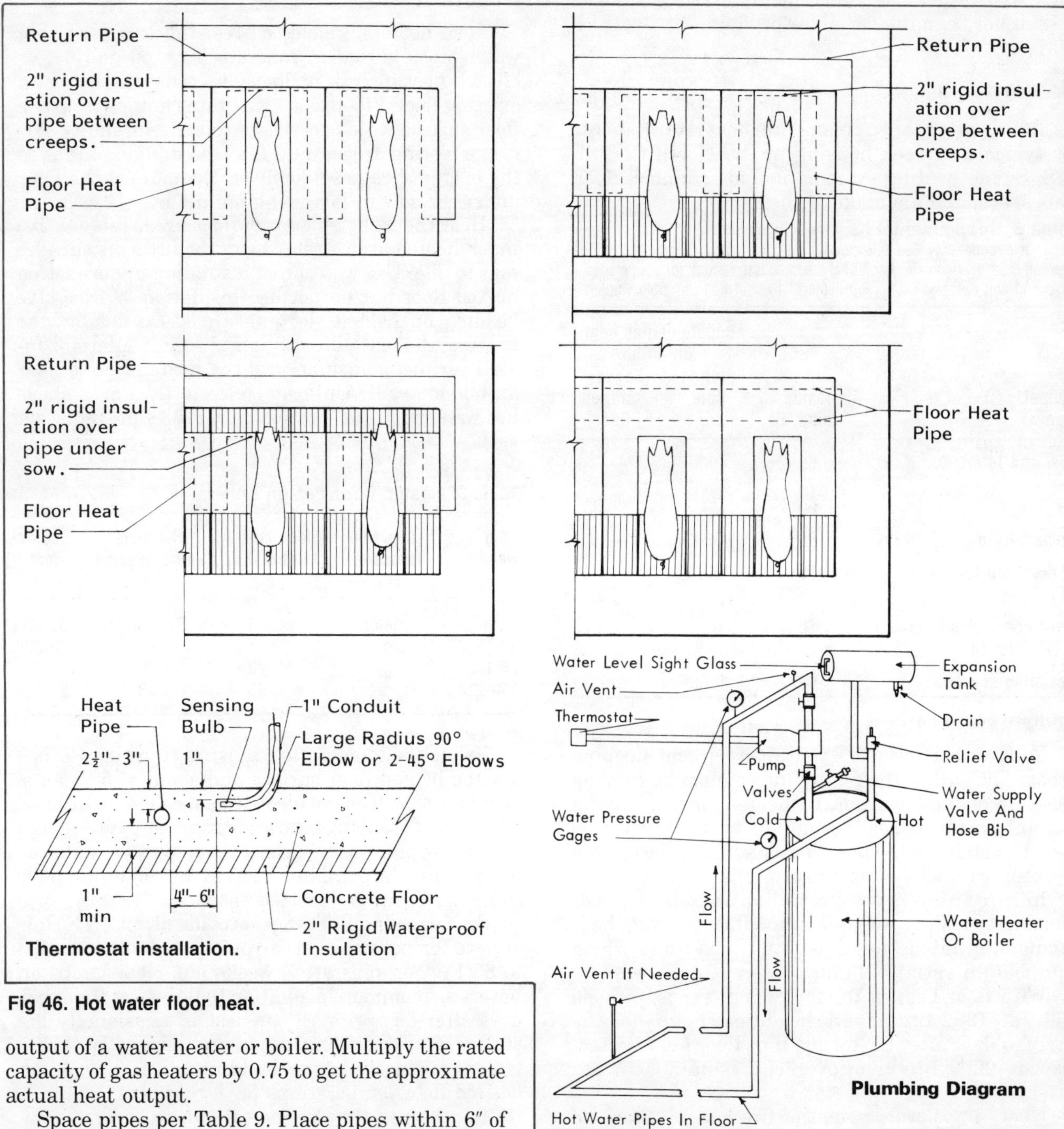

Fig 46. Hot water floor heat.

output of a water heater or boiler. Multiply the rated capacity of gas heaters by 0.75 to get the approximate actual heat output.

Space pipes per Table 9. Place pipes within 6″ of concrete slab edge and cover with 2½″-3″ of concrete. If a heated pipe passes under a farrowing sow, place insulation between sow and pipe. Pressurize the pipes with water before placing concrete, to check for leaks. Contact an experienced plumbing contractor to design and install hot water systems.

Electric heat cable usually has a lower installation cost than hot water pipes, avoids freezing problems, and permits individual control for each pen or stall. Electric heat has a higher operating cost than hot water heat.

Use cable approved for use in concrete. The most common type of heating cable is covered with polyvinylchloride (PVC) and is rated from 2 to 7 watts per linear foot. Prefabricated pads are available that have the cable already spaced in a plastic mesh to simplify installation.

Before grouting the heating cable in place, perform a continuity test to ensure the cable will operate. Embed the cable 1½″-2″ into the concrete and space uniformly. Make sure it is completely surrounded by concrete, to prevent burnout. Use a thermostat for every one to five pens or stalls. A fused switch on each pen or stall permits disconnecting them when they are empty. Electrically ground all steel stalls and waterers.

Commercial plastic heat pads are set on the floor and plugged into electric outlets. Protect electric cable from animal abuse. Thermostatic control is recommended.

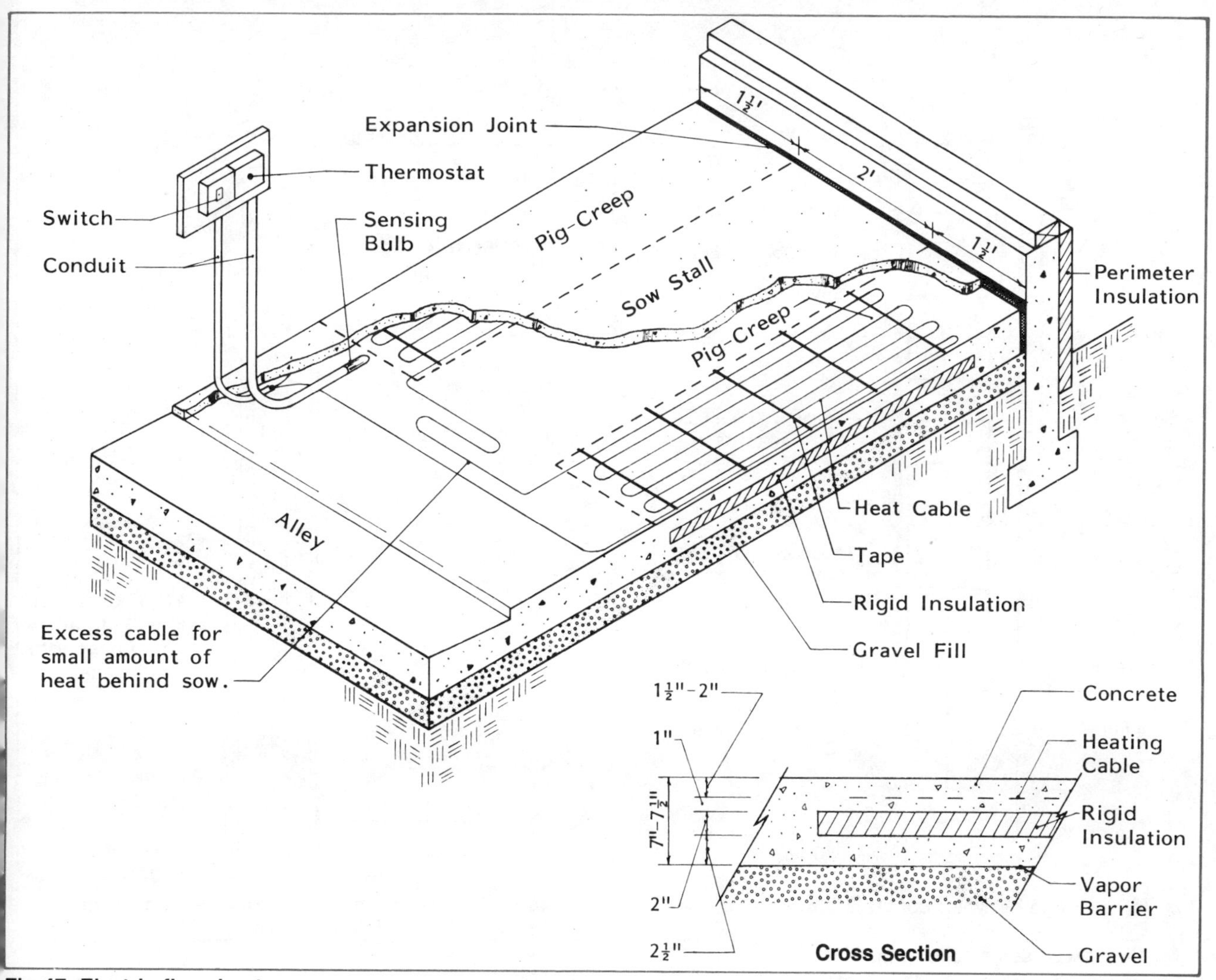

Fig 47. Electric floor heat.

Unit space heaters

Unit space heaters are located within the room and heat room air directly. Unfortunately, they recirculate dusty, wet, corrosive air through the furnace, often resulting in high maintenance. Clean and lubricate unit heaters once a month during the heating season and more often in a very dusty environment such as a growing-finishing building. If possible, place heaters to blow along the coldest wall to reduce cold drafts and radiant heat losses to the wall. With more than one heater, arrange them to create a circular air pattern within the room.

Air make-up heaters

Air make-up heaters are unvented units mounted outside the building. They heat only the incoming ventilating air, so they require less service than units which recirculate room air. Fuel flow to the burner is modulated to maintain a constant exit air temperature.

Air make-up heaters supply ventilating air, so close down the air inlet accordingly. Size air make-up heaters so they provide no more air than the winter exhaust fan.

Air make-up heaters exhaust the products of combustion into the building which makes them more efficient than vented heaters (about 90% vs. 70%). However, water vapor is one of the gases produced (about one lb water/lb propane burned), so the ventilating rate must be sufficient to remove this extra moisture. Allow 4 cfm extra ventilation for each 1000 Btu/hr of heater capacity.

Other heating systems

Solar heat may be competitive with other energy sources, depending on location and availability of energy tax credits. There are two types of solar systems:

- Passive systems have transparent or translucent panels which allow the sun to shine directly into the building.
- Active systems transport heat in air or water from a solar collector to storage and then to the animal environment.

A passive system is relatively inexpensive, but has high heat loss and moisture condensation on the poorly insulated solar panels. Although the active

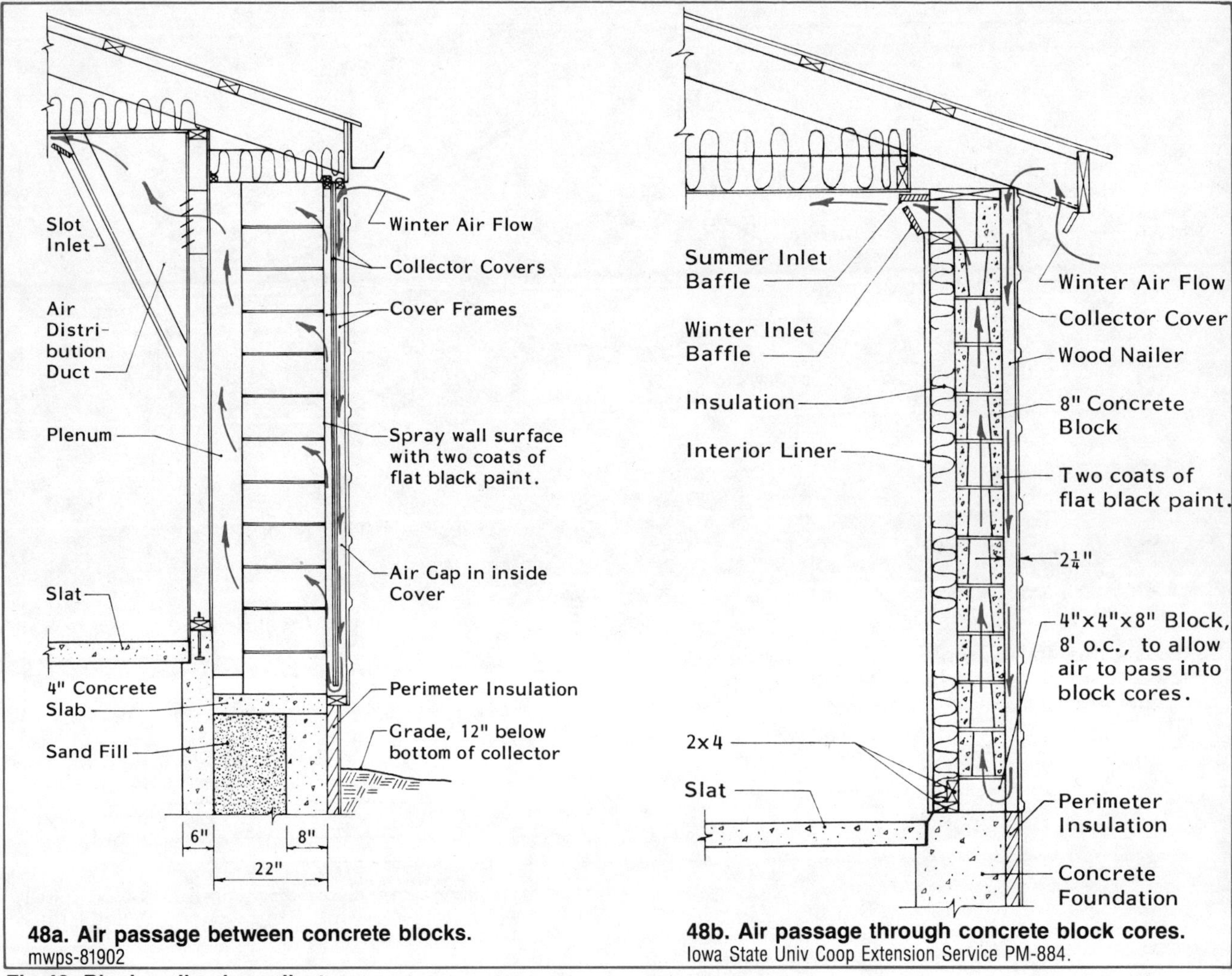

48a. Air passage between concrete blocks.
mwps-81902

48b. Air passage through concrete block cores.
Iowa State Univ Coop Extension Service PM-884.

Fig 48. Block wall solar collectors.

system is more expensive to construct, it operates more efficiently.

A block wall collector combines both collection and storage. A south-facing concrete block wall is painted black to collect and store solar heat. Ventilating air drawn past or through the sun-heated blocks gains heat before entering the building, Fig 48.

Reusing exhaust air from one livestock building to ventilate an adjacent one is not recommended. Unless the first building is overventilated, its exhaust air is nearly saturated, so it has little capacity to remove moisture from the second building. If you overventilate the first building, drafts are created and fuel savings are almost eliminated. Veterinarians and animal scientists discourage reusing ventilating air because of increased possibility of disease spread, particularly for farrowing and nursery pigs.

Heat exchangers extract heat from exhaust air which otherwise would be lost. Although this concept has a great deal of merit, dust and condensation causes equipment problems and decreased efficiency.

Earth tempering systems extract heat from the soil through pipes buried in the ground. The pipes handle cold and mild weather ventilating rates. Hot weather ventilation is conventional. Design criteria are now very limited, but research is yielding more reliable information.

Draft Control

Hovers reduce vertical drafts better than solid covers placed over slotted floors. Hovers need not be insulated—try tempered hardboard, sheet metal, or exterior plywood. Heavy clear plastic on a frame is excellent and allows you to observe the animals.

Provide enough hover space for all animals in a pen to be under it at one time—about half of a farrowing creep area or a third of a nursery pen area. Locate hovers at no more than twice the animal height and preferably next to solid pen partitions.

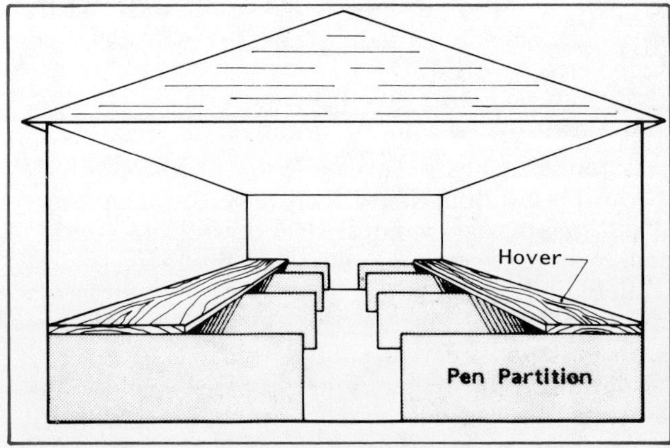

Fig 49. Hovers over animal resting areas.

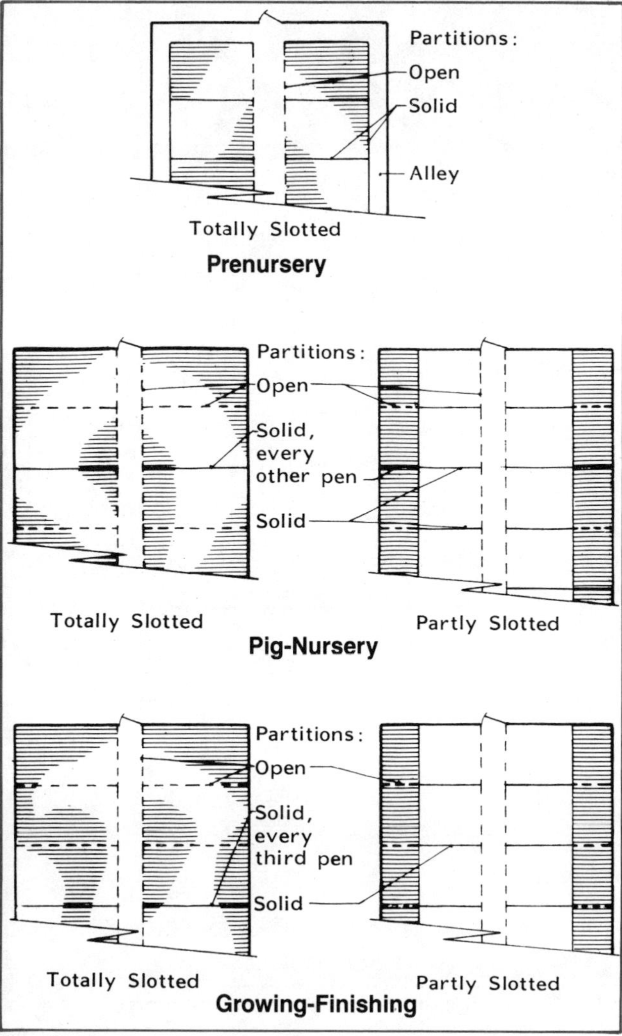

Fig 50. Draft control in environmentally controlled buildings.

Fire safety is essential for hovers with heaters. Put heaters beyond animal reach. Protect combustible materials with fire-resistant materials. Floor heat works well under hovers and reduces fire hazard.

Solid pen partitions also reduce some drafts, Fig 50.

Maintaining the Environmental Control System

Environmental control systems require periodic maintenance that is conscientious and thorough. Use the following to develop a maintenance schedule:

Every month:

- Clean fan blades and shutters. Dirty fan shutters can decrease fan efficiency by 25%. Shut off power to thermostatically controlled fans before servicing them.
- Check fans with belt drives for proper tension and correct alignment. If too tight, belts may cause excessive bearing wear; if too loose, excessive slippage and belt wear may result.
- During the heating season, remove dust from heater cooling fins and filters, and check gas jets and safety shut-off valves for proper operation.

Every 3 months:

- Make certain that shutters open and close freely. Apply a few drops of graphite (not oil or grease) to fan shutter hinges.

Every 4 months:

- Clean motors and controls. Dirty thermostats do not sense temperature changes accurately or rapidly. Dust insulates fan motors and prevents proper cooling. If dust is allowed to build up, the motor can overheat.

Every 6 months:

- Most ventilating fans have sealed bearings and do not require lubrication. However, oil other bearings with a few drops of SAE No. 10 non-detergent oil. Never overlubricate.

Every year:

- Clean and repaint chipped spots on fan housings and shutters to prevent further corrosion.
- During winter, cover hot weather fans (not cold or mild weather fans) with plastic or an insulated panel on the warm (animal) side of the fan and disconnect the power supply, or operate briefly every few weeks.
- Check slot air inlets for debris.
- Check gable and soffit air inlets for blockages.
- Check plastic baffle curtains. They can become brittle with age and require replacement.
- Check recirculation air ducts for dust accumulation.

Emergency Ventilation

Provide for emergency ventilation in all environmentally controlled buildings because of the danger of animal suffocation. This can be as simple as several manually-opened sidewall doors or as sophisticated as an electric generator system which starts automatically to power fans in case of electrical failure. Magnet-locked ventilating doors, which drop open when electrical power is cut off or room temperature rises sharply, are available commercially. Consider installing an alarm system to let you know when electrical power is off. Test your emergency ventilation and alarm systems monthly or according to manufacturers' instructions.

MANURE PIT VENTILATION

Pit ventilation reduces manure gas in the animal area, reduces odor levels, ensures more uniform air distribution, and helps warm and dry the floor. All totally and partly slotted floor buildings benefit from pit ventilation, even though manure may be removed frequently by flushing or scraping. Pit ventilation is most beneficial in winter, because it gives good air distribution at the low ventilating rates.

Pit Ventilation

There are two types of systems:

- **A duct** the length of the pit with small openings spaced to draw air uniformly from the pit.
- **An annex** outside the pit wall with a fan to draw air directly out of the pit.

Regardless of the system, allow at least 12″ clearance between the bottom of slat support beams and the manure surface. Variable-speed fans are not recommended unless the lowest rate needed is at least 40% of the fan's maximum rated capacity or unless the fan is protected with a windbreak, Fig 45.

Ducts

Ducts are expensive, but perform better than annexes. They work especially well with wire mesh or other slotted floors that have a high percentage of open area.

Ducts can be under the center alley of totally slotted buildings, Fig 51, or under the floor adjacent to the pit of partly slotted buildings, Fig 52. Select ducts and fans to ventilate at the cold weather rate. Have no point in the duct farther than 60′ from the fan.

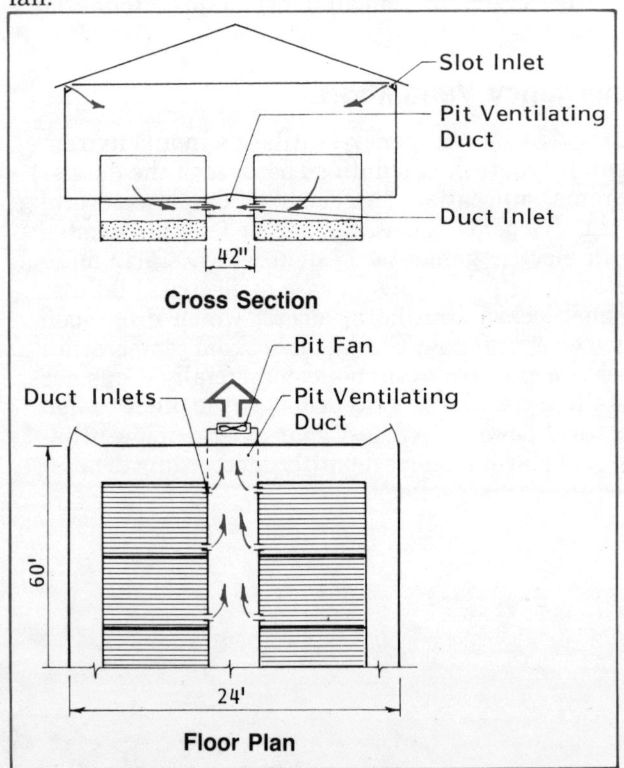

Fig 51. Pit ventilation in totally slotted floor building.

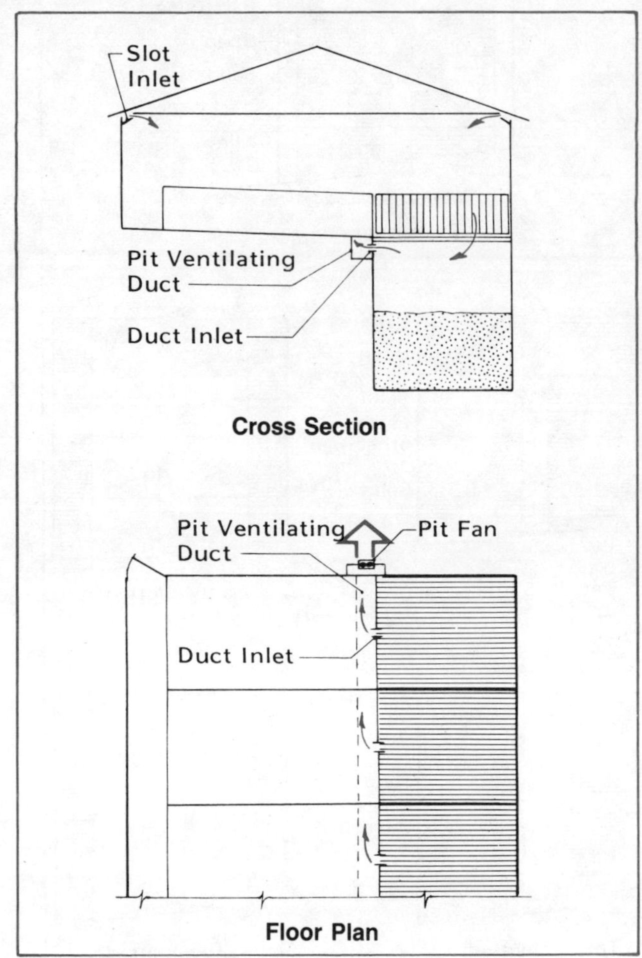

Fig 52. Pit ventilation in partly slotted building.

Use Tables 10 and 11 to determine duct and inlet opening sizes. Place a duct inlet at each pen or stall. In long buildings, place a fan at each end and/or at the center of the duct to reduce duct airflow and permit a smaller duct.

Example: Design duct pit ventilation for a 24′x60′ slotted floor, 20-stall farrowing building which has 42″ between pits, Fig 51.

1. From Table 7, pit ventilation = (20 cfm x 20 sows) = 400 cfm.
2. Select a fan for 400 cfm at ⅛″ static pressure.
3. From Table 10, 400 cfm requires a 96 in^2 duct. If the duct width is 42″, the depth has to be at least (96 in^2 ÷ 42″) = 2¼″; but 8″ could be used to simplify construction.
4. From Table 11, a per-stall ventilating rate of 20 cfm requires a duct opening of 3¾ in^2 or about a 2¼″ diameter hole.

Annexes

An annex is common in totally slotted buildings, Fig 53. It does not provide as uniform an air distribution through the floor as a duct, but it is cheaper.

Table 10. Ventilation duct sizes.
Based on a duct air velocity of 600 fpm.

Airflow rate cfm	Area in²	Minimum duct size- Inside dimensions- WxH, in.	WxH, in.	diam, in.
200	48	6x8	4x12	8
250	60	8x8	6x10	9
300	72	8x9	6x12	10
350	84	9x10	6x14	10
400	96	10x10	6x16	12
500	120	10x12	6x20	12
600	144	12x12	6x24	14
700	168	12x14	8x22	15
800	192	14x14	8x24	16
900	216	14x16	8x28	18
1000	240	16x16	8x30	18
1250	300	18x18	10x30	
1500	360	18x20	12x30	
2000	480	20x24	14x36	
2500	600	24x26	16x38	
3000	720	24x30	18x40	
3500	840	28x32	18x48	
4000	960	30x32	20x48	
5000	1200	34x36	24x50	
6000	1440	36x40	24x60	
7000	1680	40x42	28x60	
8000	1920	44x44	32x60	
9000	2160	46x48	36x60	
10000	2400	48x50	40x60	
12000	2880	54x54	48x60	
15000	3600	60x60	50x72	

Table 11. Inlets to pit ventilation ducts.
Put inlets at each pen or stall to improve room air distribution. These sizes are critical—do not vary. Based on an air velocity of 800 fpm through the opening.

Airflow cfm/pen	Opening area per pen in²	Net inlet sizes per pen HxW, in.	diam, in.
20	3.75	2x2	2¼
25	4.5	2x2¼	2½
30	5.5	2x2¾	2¾
35	6.25	2x3¼	3
40	7.25	2x3¾	3½
60	10.75	3x3¾	3¾
80	14.5	3x5	
100	18	4x4½	
125	22.5	4x5¾	
150	27	4x6¾	
200	36	4x9	
250	45	4x11¼	
300	54	2-4x6¾	
350	63	2-4x8	
400	72	2-4x9	
450	81	2-4x10¼	
500	90	2-4x11¼	
550	99	2-4x12½	
600	108	2-4x13½	
700	126	2-4x15¾	

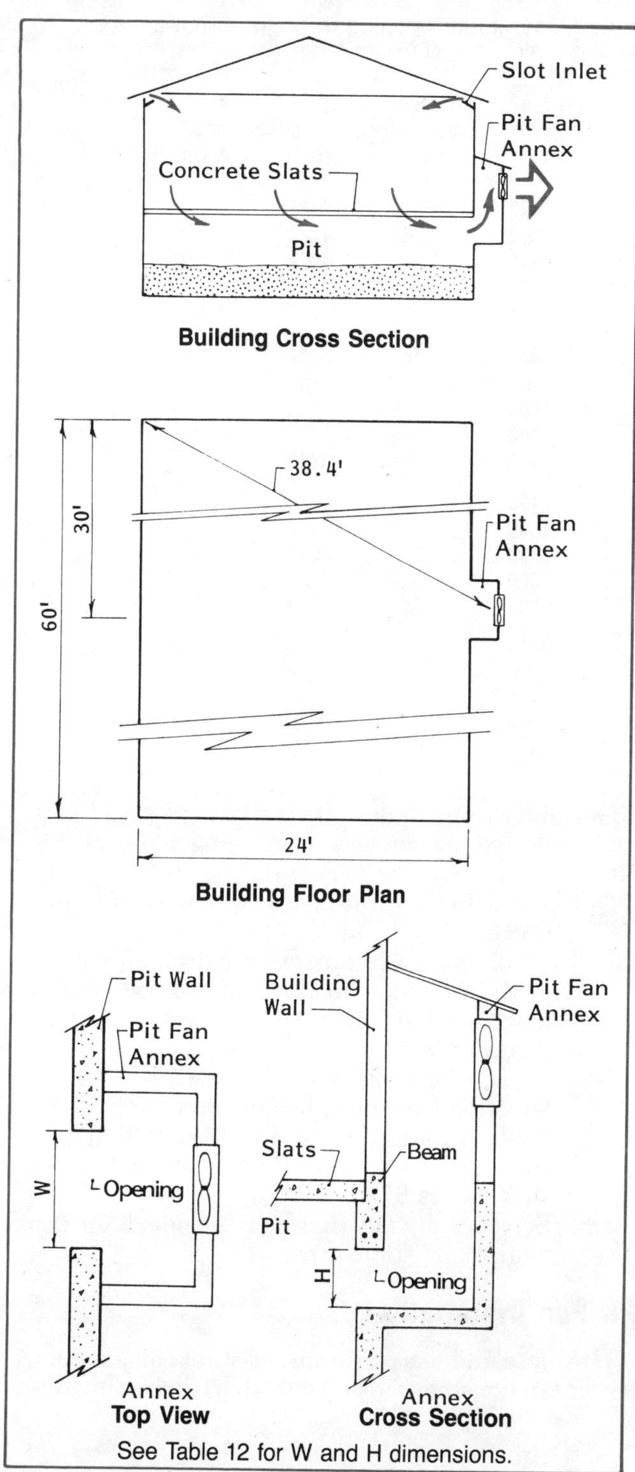

Fig 53. Exhaust pit fan annex.

Annexes work best for concrete slats with only 15%-20% open area because of the duct effect created by the slats. Air distribution under wire or metal flooring with 50%-60% open area is poor.

Design annex pit ventilation to supply at least the cold weather rate but no more than the mild weather rate. Table 12 shows the minimum pit wall opening for various fan capacities. Larger openings will not significantly affect performance. Locate annexes so no point in the pit is farther than 50′ from an annex.

Table 12. Minimum opening through pit fan annex.
Based on an air velocity of 800 fpm through the opening.

Pit fan capacity cfm	Opening area in²	Inside dimensions WxH, in.	diam, in.
400	72	4x18	10
500	90	4x23	12
600	108	4x27	12
700	126	4x32	
800	144	4x36	
900	162	6x27	
1000	180	6x30	
1100	198	6x33	
1200	216	6x36	
1300	234	6x39	
1400	252	6x42	
1500	270	8x34	
1600	288	8x36	
1800	324	8x42	
2000	360	12x30	
2500	450	12x38	
3000	540	12x45	
3500	630	18x36	
4000	720	18x40	
5000	900	24x38	

Example: Design annex pit ventilation for a 24'x60' totally slotted pig-nursery for housing 400 pigs, Fig 53.

1. From Table 7, the cold weather rate for nursery pigs is 3 cfm. The pit ventilation is:
   (3 cfm x 400 nursery pigs) = 1200 cfm.
2. With one annex in the center of the winter downwind sidewall, the maximum air-pull distance is:
   Square Root $[(60' \div 2)^2 + (24')^2] = 38.4'$, which is less than the 50' limit.
3. Table 12 indicates a pit opening of 216 in² into the annex for 1200 cfm. Minimum opening for this rate is 6" x 36".
4. Select a pit fan that can provide 1200 cfm against ⅛" static pressure.

Pit Fan Installation

Insulate and use corrosion resistant construction for pit fan housings. To prevent "short-circuiting" between fans, put each pit fan in its own duct or annex. Shutters are not necessary on continuous fans.

If odor around the building is a concern, discharge ventilating air high on the sidewall, which disperses odors faster than if discharged at ground level. Dispersion can be improved by adding a chimney to exhaust even higher, Fig 54. Construct the chimney about 50% larger than the fan diameter to limit chimney air velocity to 1000 fpm or less. Put a drain hole at the base of the chimney for condensation or rainwater. Protect metal siding from corrosion with exterior plywood next to the building. Insulate the chimney liner to reduce frost formation and snow buildup.

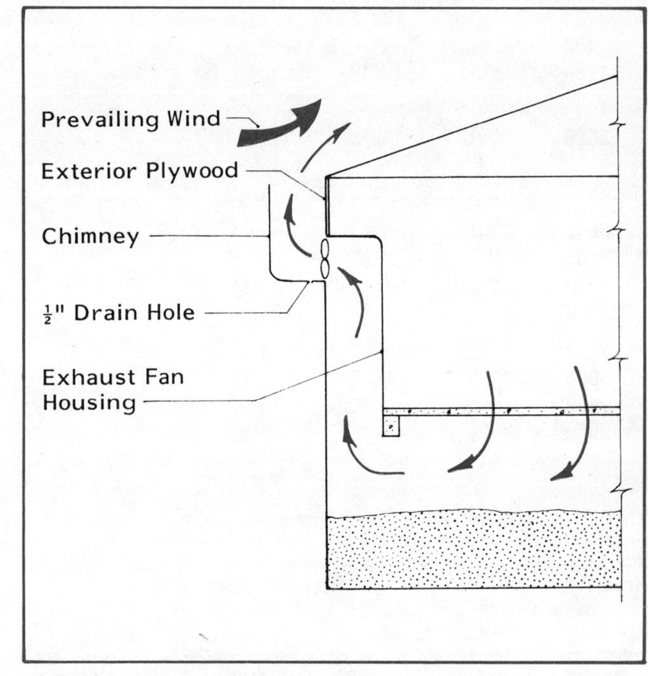

Fig 54. Chimney exhaust vents.
Locate vent on upwind side of building.

Chimneys also protect variable-speed fans from wind, allowing a more uniform ventilating rate. Exhausted materials may collect on metal roofs, causing corrosion. Check roofing around the chimney each year and retouch with a zinc base paint. Other solutions: extend the chimney 5'-6' above the roofing for better dispersion or use asphalt base roofing.

COOLING SYSTEMS

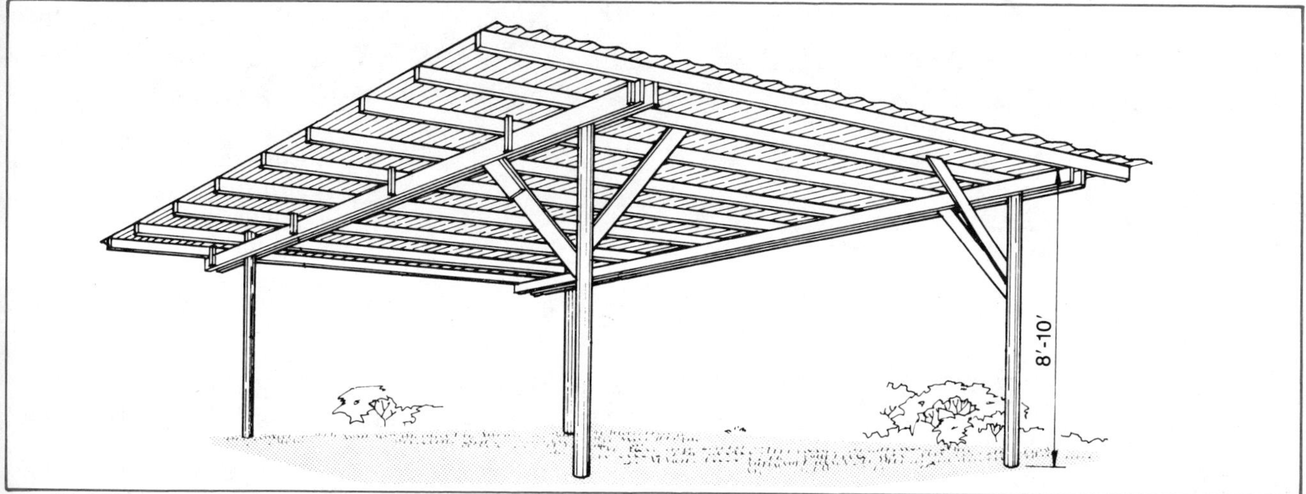

Fig 55. Livestock shade.
Provides both shade and a cool ground surface to lie on.

Hot weather reduces feed intake of growing-finishing swine and decreases the reproductive capability of breeding stock. Swine over 50 lb feel heat stress at about 70 F, and temperatures above 85 F may cause substantial losses unless they are kept cool.

Shade Cooling

Shades are effective for cooling livestock in pastures and outside lots. Build shades at least 8′ high (preferably 10′), with their long axis oriented east-west. High shades increase exposure to the "cool" northern sky, which increases radiant heat loss from the animal.

Shades with straw roofs are most effective because they insulate as well as reflect sunlight. Aluminum or bright galvanized steel roofs are also good. Painting the upper surface white and the lower black adds a 10% benefit. Snowfencing, a common shade material, is about half as effective as straw or painted metal.

Water Cooling

Drinking Water

Animals drink a lot of water in hot weather so their evaporative heat loss system can keep them cool. Cool drinking water provides the most relief.

Wet-Skin Cooling

Substantial cooling is possible by wetting the animal's skin, then allowing the moisture to evaporate. A slight breeze across the animal increases the evaporation rate and improves cooling even more.

Pasture wallows provide wet-skin cooling. Shaded wallows are more effective because they reduce solar heat and the water remains cooler.

The best sprinkler systems wet the animal, then allow it to dry. A thermostat- and timer-controlled sprinkler wets the pigs for 2 to 3 minutes out of each hour that temperatures are above 85 F. Tables 13 and 14 suggest water line and nozzle sizes. Select non-corrosive nozzles designed to furnish a solid cone of water droplets, **not** a fog. A fogger cools the air; a sprinkler cools the animal directly.

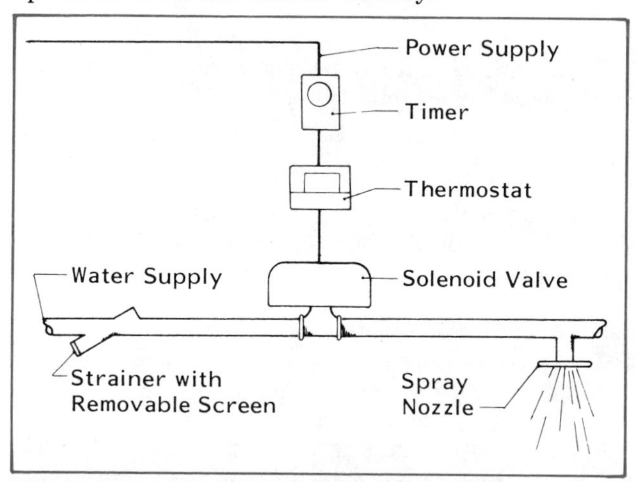

Fig 56. Sprinkler with solenoid valve control.

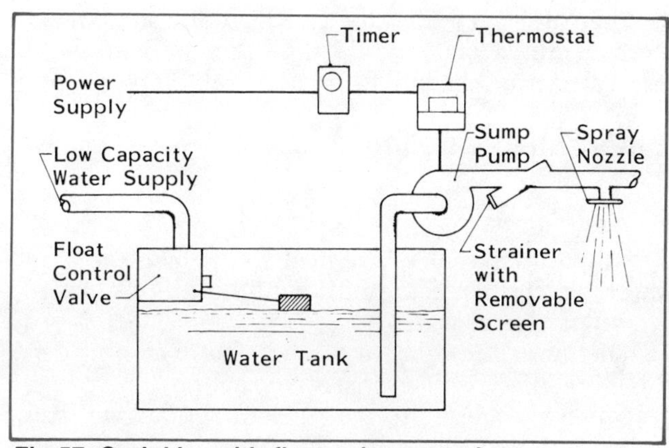

Fig 57. Sprinkler with float valve control.
A float valve and tank are often less costly and less complicated than a solenoid valve and manifold piping. Select a sump pump for at least 30 psi pressure.

Table 13. Nozzles for swine sprinklers.
One nozzle per pen about 6′ above floor.

Pigs per pen	Nozzle size
10	0.45 gpm
20	0.90 gpm
30	1.35 gpm

Table 14. Water lines for swine sprinklers.
Plastic pipe: Based on 5 psi/100′ maximum pressure drop and 4 feet/second maximum velocity.

Pipe size in. o.d.	Maximum flow, gpm
½	3
¾	6
1	10
1¼	18
1½	25
2	40
2½	50
3	90
4	140

Ventilation Cooling

Hot weather ventilation cools swine by:

- Replacing hot air with cooler air.
- Taking high-humidity respirated air away from the animals.
- Increasing air velocity around the animals (wind chill).
- Using heat from the air to evaporate moisture from the building surfaces, creating an evaporative cooling effect.

Hot weather ventilation maintains inside temperatures no more than 3 F higher than outside conditions. Hot weather rates in Table 7 are about twice as high as needed to remove animal heat from a well-insulated building. They are designed for high air velocities around the animals. At least 150 fpm is necessary for appreciable hot weather cooling of larger animals. With some systems, you can direct fresh air onto the animals. One half of the total hot weather rate can be obtained with circulation fans to increase air velocity, while the other half must be exhausted from the building to control temperature.

Evaporative Cooling

Principles

Evaporative coolers use heat from air to vaporize water, which increases relative humidity but lowers the air temperature.

The lower the relative humidity of incoming air, the more effective is evaporative cooling. Therefore, evaporative coolers are more useful in dry western states, but they are still effective in the Midwest. Humidity drops as air temperature rises and is usually lowest during the hottest part of the day. Expected temperature drop during midsummer is shown in Fig 58.

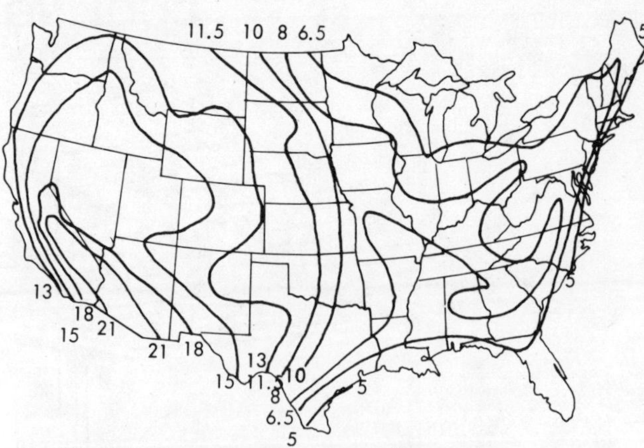

Fig 58. Effectiveness of evaporative coolers.
Average room temperature drop (F) from maximum outdoor July temperature.

Commercial evaporative coolers are available for livestock buildings. Air is drawn through wet pads or foggers into the animal area. Most units pump water from a sump over a fibrous pad. One commercial unit rotates a drum-mounted pad through a pan of water.

Provide airflow through the evaporative unit of at least one half the hot weather rate, Table 7. On hot, humid days evaporative cooling will not be very effective, so provide additional airflow around the animals. Install circulation fans sized at one-half the hot weather rate, or make the ventilating system able to supply the full hot weather rate on those days.

Design

Pad area (ft^2) for a wall evaporative cooler, Fig 59, is the airflow through the pad (cfm) divided by 150 for aspen pads, or by 250 for cellulose pads. Provide insulated doors or covers to close off the pads during winter.

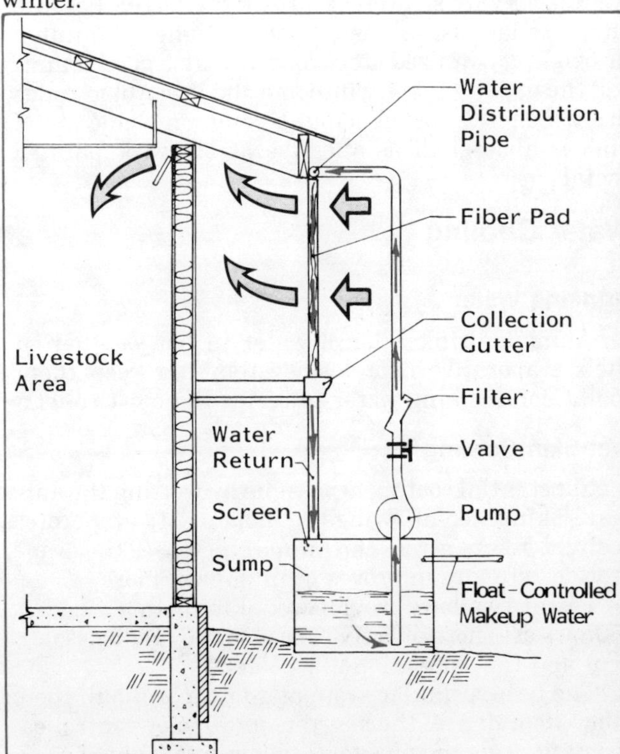

Fig 59. Evaporative cooling system.
Ventilating exhaust fans pull hot air through wet fiber pads. Air heat evaporates the water and lowers air temperatures.

Supply water to the pads with an overhead system. Rigid plastic pipes or open rain gutters with spaced holes allow water to drip uniformly over the pads. Pipe size, hole size, and spacing depend on water flow-rate and must be individualy sized for each system. For best evaporative efficiency, much more water is supplied than is used. To conserve unevaporated water, reuse it.

Install a water sump of at least ½ gal/ft² of aspen pad or ¾ gal/ft² of cellulose pad. Provide a flow-rate of at least ⅓ gpm/linear ft of aspen pad or ½ gpm/linear ft of cellulose pad to the distribution pipe over the pads. A gutter beneath the pads, sloped at 1"/10', collects and conveys unevaporated water back to the sump. Control make-up water to the sump with a shut-off float valve. Up to 1 gpm/100 ft² of pad evaporates on hot, dry days.

Protect the water distribution system from insects and debris. Screen recirculated water before it returns to the sump. Install a filter between the pump and the distribution pipe or gutter.

Set the thermostat to begin wetting pads at about 85 F. To reduce algae growth, wire the pump to stop several minutes before the fans so the pads dry out after each use.

Maintenance

Most pads are woven aspen fibers and must be replaced annually. Cellulose pads with a life of about 5 years are available with some units. Pads are mounted either on the sidewalls, endwalls, or roof. Wall-mounted units are easier to maintain than roof units. If the pad settles, fill in the opening so air does not short-circuit.

Hose pads off at least every other month to wash away dust and sediment. Control algae buildup in the water with a copper sulfate solution. Light-tight enclosures around pads and water sump helps control algae. Water is constantly evaporated, so salts and other impurities build up. Bleed off 5%-10% of the water continuously to flush salts from the system as they are formed or flush out the entire system every month.

Air Conditioning Systems

Livestock rooms are seldom air conditioned because it is expensive. Air conditioners cannot reuse cooled air from livestock buildings because of corrosive gases and dust, so they continually cool hot air in a one-pass process.

Zone Cooling

In hot weather, a hog loses 60%-70% of its heat through evaporation from the respiratory tract and convection from the skin surface, so it helps to zone cool the area around its head. Zone cooling is generally used for stall- or tether-restrained animals and occasionally for animals in individual pens. In farrowing buildings, zone cooling cools the sow while allowing higher temperatures in the pig creep.

Zone cooling does not satisfy all of the hot weather ventilation needs. You still need to provide one-half of the hot weather ventilation rate with conventional ventilation.

A zone cooling system has a main air duct and downspouts (or drop ducts) located as needed for the animals, Fig 60. Locate downspouts close to the animals' head; if spouts are too far away, the cooled air mixes with room air, decreasing its effectiveness. If the outlet is within the animals' reach, make it pig-proof. Provide dampers to close downspouts when stalls are empty.

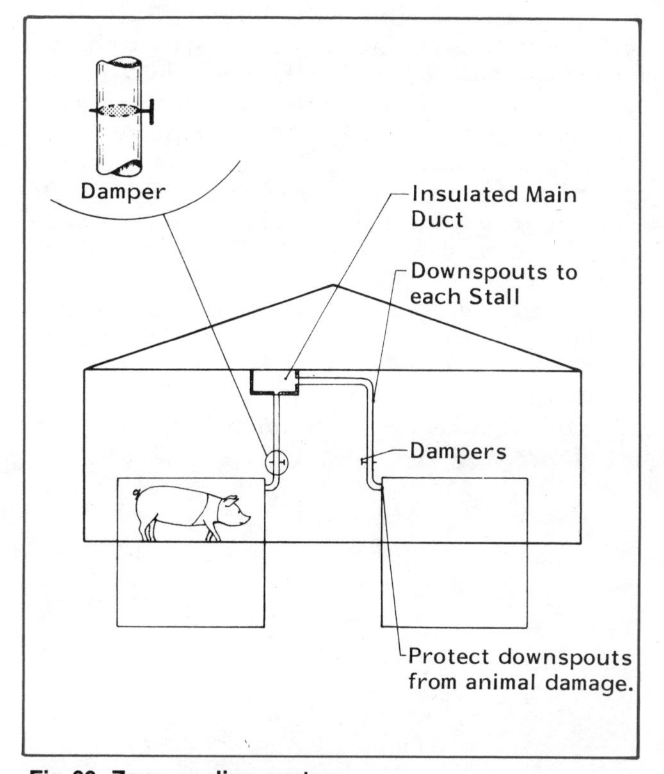

Fig 60. Zone cooling system.
Zone cooling provides only part of total ventilation but is very effective, because high velocity air is ducted directly to the animal.

Table 15 recommends ventilation rates for zone cooling systems. Tables 10 and 16 recommend main duct and downspout sizes. The main duct may be slightly larger, but not smaller. Downspout sizes are critical, but downspouts with dampers can be slightly larger. Insulate ducts to prevent excessive heat gain and condensation.

Table 15. Airflow for zone cooling swine.
In addition to zone cooling provide at least one-half of the hot weather rate in Table 7 with conventional ventilation.

	Airflow for:		
Type of animal	Uncooled air	Evaporative cooled air	Conditioned air
	- - - - - -cfm/animal- - - - - -		
Farrowing sow	70	40	35
Gestating sow	35	20	15
Boar	55	30	20

Example: Determine the airflow and duct size requirements of the three zone cooling systems for a 20-sow farrowing building.

In addition to the zone cooling, ventilate the room with a conventional ventilating system. Size at one-half the hot weather rate in Table 7: (250 cfm x 20 sows) = 5000 cfm.

Uncooled air: From Table 15, recommended airflow is 70 cfm in each downspout and (70 cfm x 20 sows) = 1400 cfm in the main duct. From Tables 16 and 10, 70 cfm requires a 4" diameter downspout and 1400 cfm requires a 18"x20" main duct.

Evaporative cooling: From Table 15, recommended airflow is 40 cfm in each downspout and (40 cfm x 20 sows) = 800 cfm in the main duct. From Tables 16 and 10, 40 cfm requires a 3" diameter downspout and 800 cfm requires a 14"x14" main duct.

Air conditioning: From Table 15, recommended airflow is 35 cfm in each downspout and (35 cfm x 20 sows) = 700 cfm in the main duct. From Tables 16 and 10, a 3" diameter downspout and a 12"x14" main duct is required.

Size a zone air conditioning system at about one ton of cooling capacity per each 100 cfm (e.g. about one-third ton per sow and litter). Some of the cooling effect is lost because the cooled air gains heat as it passes through the duct to the animal. Insulate the duct to at least R = 6.

Table 16. Downspout sizes for zone cooling.
Sized for 800 fpm air velocity.

Airflow cfm	Area in^2	Inside dimensions WxH, in.	diam, in.
10	1.8	1½x1½	1½
15	2.7	1¾x1¾	2
20	3.6	2x2	2½
25	4.5	2x2¼	2½
30	5.4	2x2¾	3
35	6.3	2x3¼	3
40	7.2	2x3¾	3
45	8.1	3x2¾	4
50	9.0	3x3	4
55	9.9	3x3½	4
60	10.8	3x3¾	4
65	11.7	3x4	4
70	12.6	3x4¼	4
75	13.5	3x4½	4
100	18	4x4½	6
125	22.5	4x5¾	6
150	27	4x6¾	6

NATURAL VENTILATION

Natural ventilation works best for swine over 75 lb. It is suitable for small pigs on bedded floors or in well-insulated modified open-front buildings with zone heat.

Three types of buildings are naturally ventilated, Fig 61:

- **Small housing units:** individual farrowing buildings and portable shelters. Most have an outside exercise lot.
- **Open-front buildings:** solid floor with heat or bedding and outside lot.
- **Modified open-front (MOF) buildings:** much of the sidewall is opened in summer and closed in winter. Well-insulated MOF buildings work well for finishing and gestating swine.

Principles of Natural Ventilation

The difference between inside and outside wind pressure and temperature moves air through the building. Natural ventilation of gable buildings works best with small eave and ridge openings, large sidewall openings, and no ceiling.

Winter ventilation: wind across the open ridge creates suction which draws warm, moist air out through the ridge and fresh air in through the eave openings. The downwind eave openings may occasionally act as air outlets. Some ventilation occurs even on calm days because warm air rises, but this chimney effect accounts for only about 10% of winter ventilation.

Summer ventilation: large openings (typically one-third to one-half of the sidewalls) in each sidewall allow a cross flow of air. The ridge opening has little effect in summer.

Site Selection

Trees, tall silos, grain bins, and other structures disturb airflow around adjacent buildings. Locate naturally ventilated buildings on high ground to expose them to wind and at least 50′ (in any direction) from other structures and trees. Trees or structures affect both summer and winter ventilation for 5 to 10 times their height downwind. If naturally ventilated buildings are in a building complex, place them on the west or south side of mechanically ventilated buildings, Fig 62.

Building Orientation

Construct open-front buildings with an east-west long axis for best summer shade and winter sun penetration. The greater solar heat load on north-south oriented roofs is not a major concern for well-insulated MOF buildings.

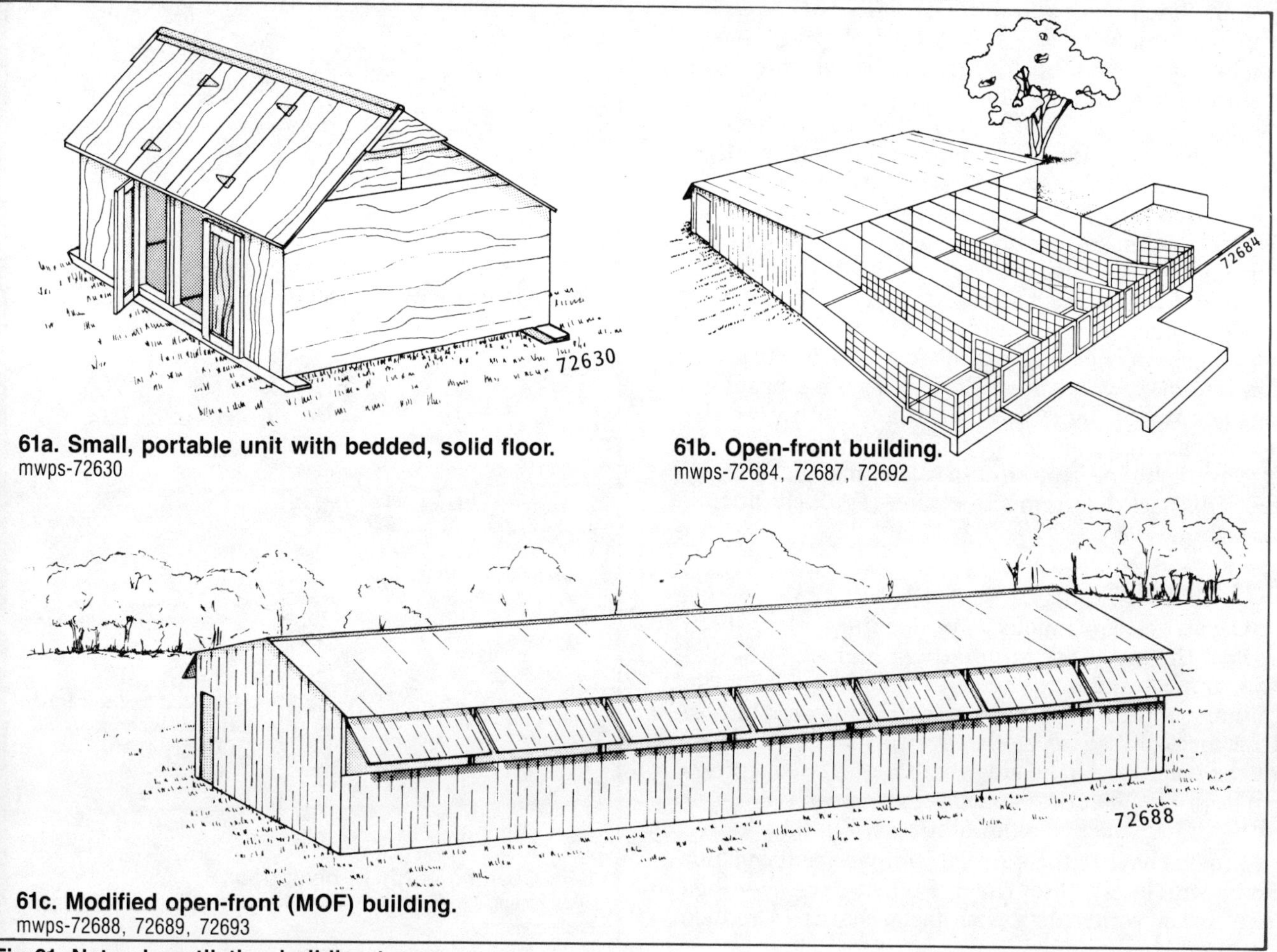

61a. Small, portable unit with bedded, solid floor.
mwps-72630

61b. Open-front building.
mwps-72684, 72687, 72692

61c. Modified open-front (MOF) building.
mwps-72688, 72689, 72693

Fig 61. Natural ventilation building types.

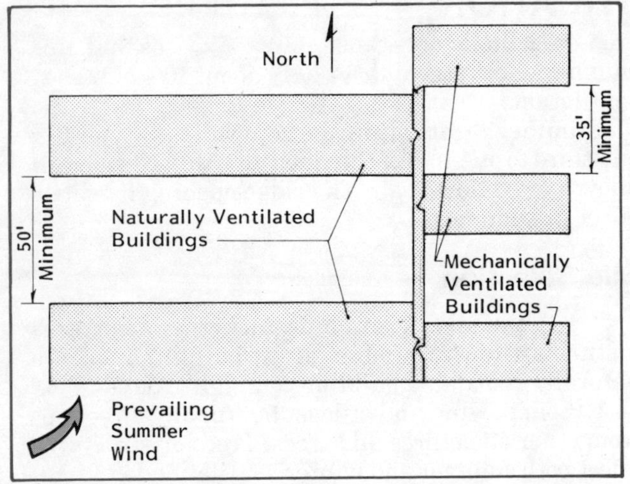

Fig 62. Layout of naturally and mechanically ventilated buildings.
Locate naturally ventilated structures on the west or south side so mechanically ventilated buildings do not interfere with prevailing windflow.

Design

Ridge Openings

In the Midwest, size ridge openings at 5 to 6 in²/finishing pig and 7 in²/gestating sow. Make the ridge opening at least 3″ wide to prevent frost buildup. If swine are uniformly distributed in the building at recommended space allowances, use Table 17 to determine opening sizes.

For open-front buildings, construct the ridge opening by removing the ridge cap and covering trusses with flashing, Fig 63a. Provide closures on MOF ridge openings or use commercial ridge ventilators with adjustable openings, so airflow can be reduced during cold periods, Fig 63b. With adjustable closures, increase opening width 50% or more to improve airflow on calm days.

Air exiting through a properly sized ridge opening prevents most precipitation from entering. Raised ridge caps are not recommended—they disturb airflow and may even trap snow. If birds are a problem, consider installing 3/4″x3/4″ hardware cloth over the ridge openings. Screened ridge openings are very susceptible to freezing shut in cold climates. You may need to knock ice from the screen regularly during cold spells.

Eave Openings

Construct continuous eave openings along both sides of the building. Locate them high on the sidewall so incoming air is warmed before reaching the animals. Screen them with 1/2″x1/2″ hardware cloth to keep birds out. Size **each** eave to have at least as much open area as the ridge opening. Eave openings can be provided at the open spaces between trusses or with slightly opened sidewall doors, Fig 64.

Provide eave baffles so airflow can be reduced, but never completely close them. Protect eave openings from direct wind gusts with facia boards to reduce drafts on the animals.

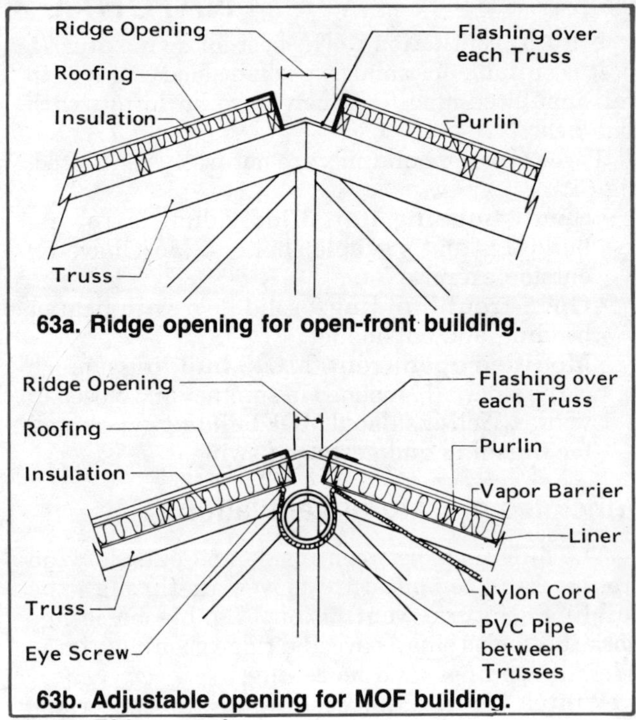

63a. Ridge opening for open-front building.

63b. Adjustable opening for MOF building.

Fig 63. Ridge openings.
Avoid exposed purlins deeper than 4″. Smooth liners under the purlins are ideal.

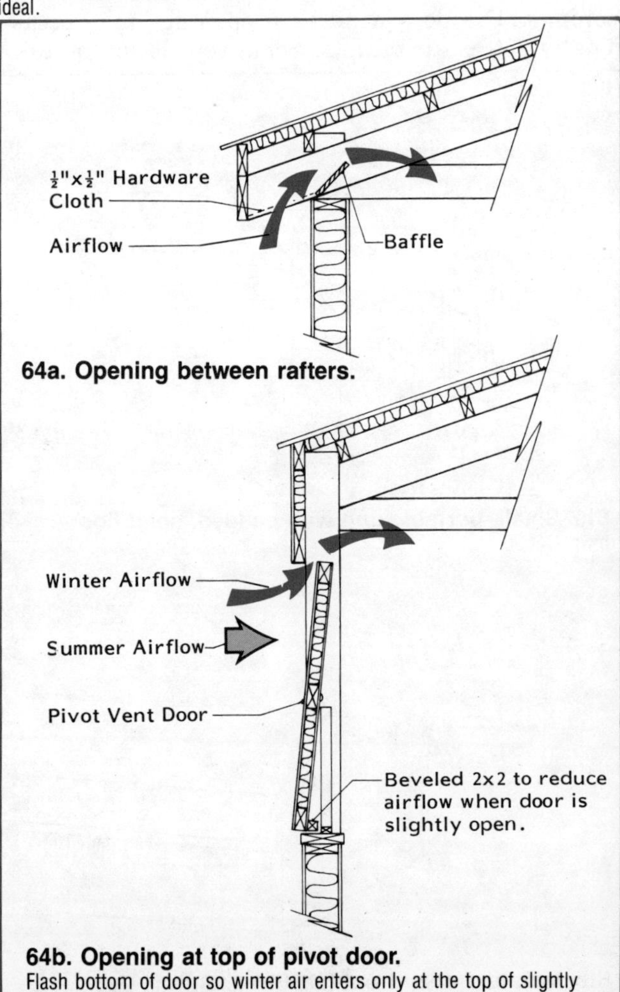

64a. Opening between rafters.

64b. Opening at top of pivot door.
Flash bottom of door so winter air enters only at the top of slightly opened doors.

Fig 64. Typical eave openings.

Sidewall Openings

Several types of summer vent doors work well, including pivot doors, top- or bottom-opening doors, and plastic curtains, Fig 64b and 65. See Table 17 to size. Construct vent doors to open to a full horizontal position so none of the opening is constricted, or increase door size so the recommended open area is still provided.

Example: Design the natural ventilation openings for a 30′ wide gable roof swine finishing building.

1. From Table 17a, **each** continuous eave inlet is 1.5″ wide (HI).
2. Assume a 3″ wide ridge opening (WO). Divide the table constant by the ridge opening width to get the length of 3″ opening required for each 16′ of building length (LO). From Table 17a, the table constant (TC) = 24.
   LO = 24 ÷ 3″ = 8′
   (i.e. 8′ of opening for each 16′ of building length)
3. From Table 17b, install a continuous 42″ wide opening along **each** sidewall.

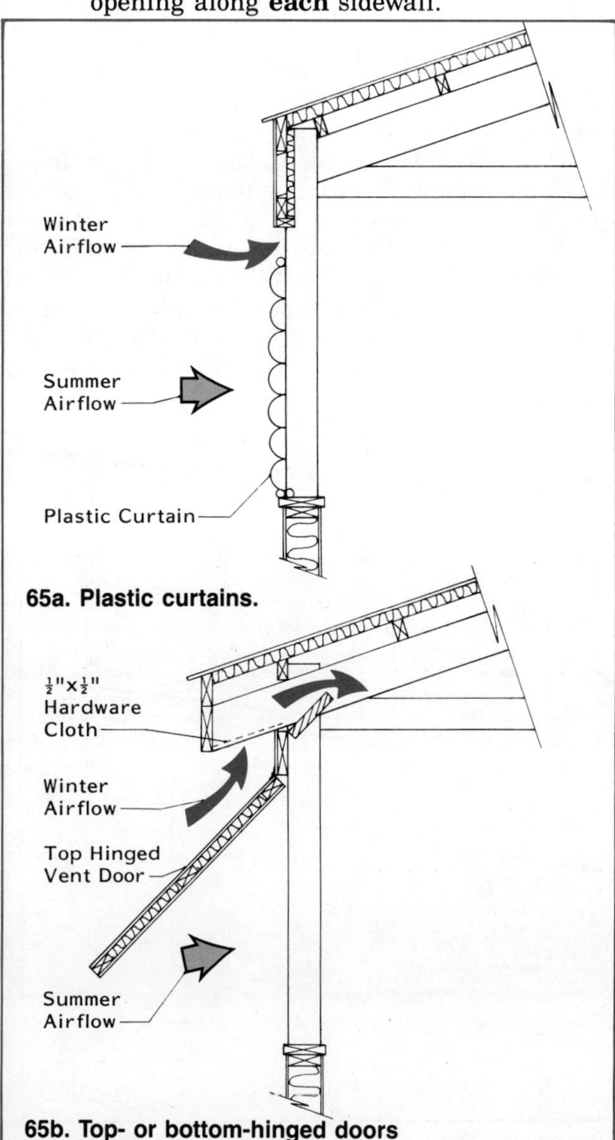

65a. Plastic curtains.

65b. Top- or bottom-hinged doors

Fig 65. Sidewall openings.
See also Fig 64b.

Table 17. Natural ventilation openings.

17a. Slot widths for cold and mild weather ventilation.
Install a continuous **winter air inlet** along each side of the building. Make **each** inlet the height (HI) in this table.

Make **winter air outlets** at least 3″ wide to prevent them from freezing shut. It is usually necessary to install several shorter vents that are 3″ or wider, instead of a continuous narrow opening. Provide an outlet at least 16′ o.c. To determine the length of the outlet opening required per 16′ of building length (LO), divide the table constant (TC) by the outlet opening width (WO = 3″ or wider).

Building width	Winter air inlet height (HI)		Winter air outlet constant (TC)	
ft	Finishing in.	Gestation in.	Finishing —	Gestation —
10-15	0.75	0.50	12	8
16-20	1.00	0.75	16	12
21-25	1.25	1.00	20	16
26-30	1.50	1.25	24	20
31-35	1.75	1.50	28	24
36-40	2.00	1.50	32	24

17b. Sidewall openings for warm weather ventilation.
Heights given are for one opening the entire building length. Provide an opening in **each** sidewall.

Building width	Gable roof	Monoslope roof	
ft	Each sidewall in.	Backwall in.	Frontwall in.
10-15	24	12	36
16-20	30	16	48
21-25	36	18	60
26-30	42	24	66
31-35	48	—	—
36-40	60	—	—

Winter Vent Openings:

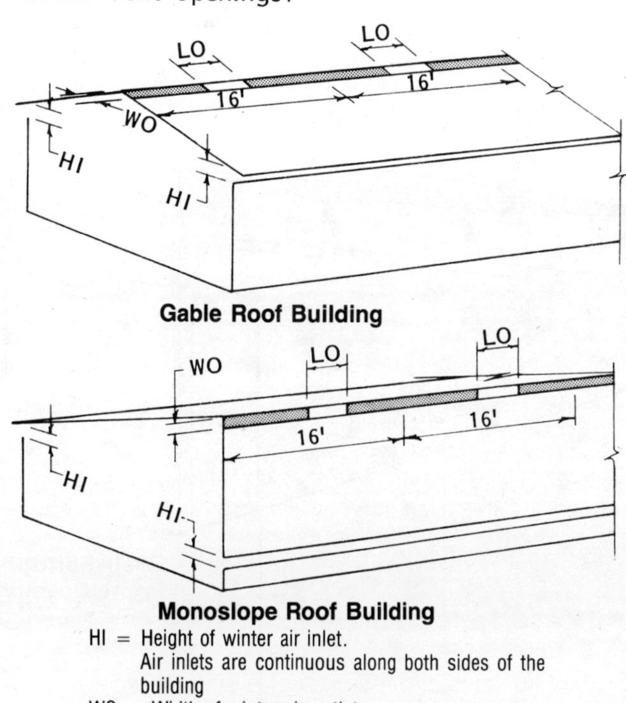

Gable Roof Building

Monoslope Roof Building

HI = Height of winter air inlet.
Air inlets are continuous along both sides of the building
WO = Width of winter air outlets.
LO = Length of air outlet per 16′ of building length.
LO = TC/WO

Roof

The steeper and smoother the roof underside, the better the airflow. Avoid roof slopes less than 3/12 for gable roofs and 2/12 for monoslope roofs. Avoid exposed purlins deeper than 4″. See section on insulation for recommended insulation values.

Management

Adjust vent openings several times a day during changing weather to avoid wide temperature fluctuations. Install a winch and cable system to adjust several doors from one convenient location. Consider thermostatically controlled electric winches for automatic door adjustment. Locate the controller at the center of building and connect main cables from both directions. This puts less stress on the controller supports than pulling from one direction. Use prestretched cable to minimize future adjustments. During cool weather, disconnect unwanted doors and control only the doors needed at that time.

Standby power units provide protection in case of power failure. Some automatic winches operate off a trickle charged battery, so door control is provided after a power failure.

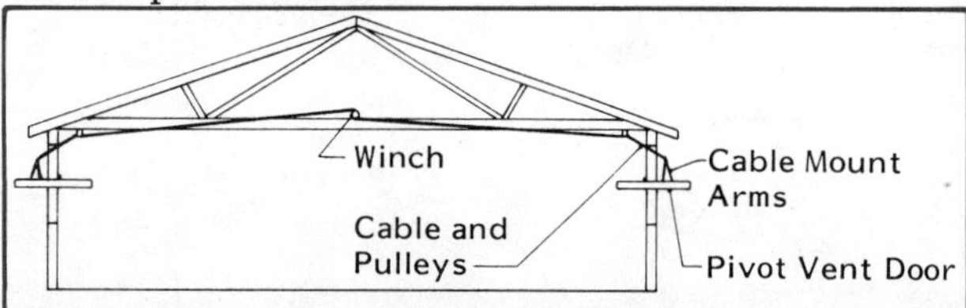

Fig 66. Cable controlled pivot doors
Mount cables on outside arms so doors can be opened to a full horizontal position. Hinge pivot doors just above center so they close when cable tension is released.

MOF Buildings

In cold weather, adjust only the eave doors to maintain desired room air conditions. Keep the ridge fully opened except during very cold periods or blowing snow. Keep the large, sidewall vent doors closed unless they are used for the eave openings.

As the weather warms, gradually open the downwind vent doors first. When downwind doors are about half open, start opening the upwind doors. Completely open all the vent doors in hot weather. Reverse this sequence as temperature decreases.

Open-Front Gable Buildings

Open-fronts have limited ventilation control, and animals must adjust to rapidly changing temperatures. Open the back sidewall doors in spring and close in fall; close eave vents when temperatures drop below freezing. Ridge vents are seldom closed, regardless of the weather.

Monoslope Buildings

Monoslopes have no ridge so both winter and summer ventilation are accomplished by a cross flow of air.

Open-front monoslopes are operated like open-front gable buildings. Orient so the open side catches prevailing summer wind while the back side blocks prevailing winter wind. Close the back wall openings in winter and open them in summer.

MOF monoslopes have large, adjustable openings in the front (high) and the back sidewalls. The back sidewall opening is usually an insulated tilt door and the front sidewall is a plastic fabric curtain, a fiberglass panel, or an insulated tilt door in cold climates. These buildings also have small, adjustable

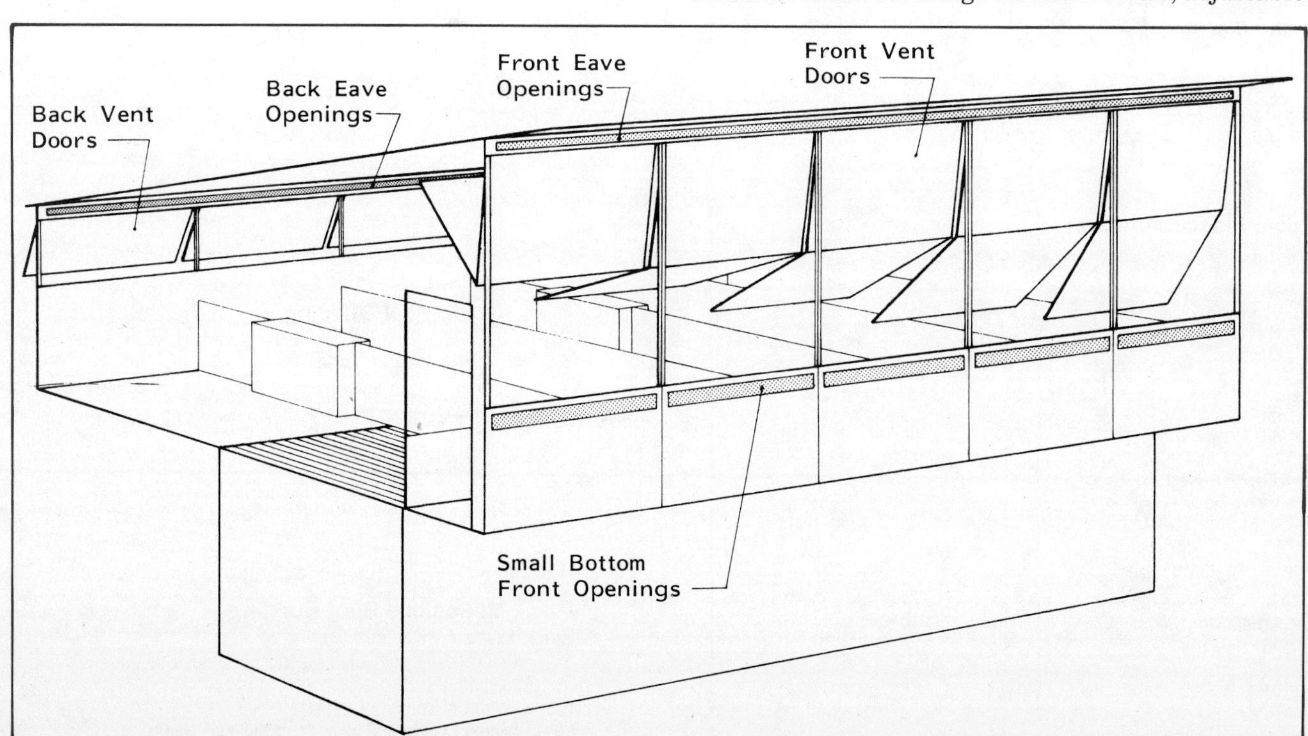

Fig 67. Modified open-front (MOF) monoslope building.

openings along the eaves of both sidewalls and along the bottom of the front sidewall, Fig 67. Size the small openings the same as the eave openings of gable roof buildings, Table 17.

When outside temperature gets below 20 F, close the back eave openings. Ventilating air enters through the small bottom front opening and moves 10′-15′ back into the animal area. As the air is warmed by the animals, it rises and travels up the sloped roof to exit through the front eave opening. The back sleeping area is not ventilated directly during extremely cold periods.

As outside temperature rises above 20 F, start opening the back eave openings and eventually the large front doors. In summer, open all the vent doors.

Controlling Drafts

Wind against a building endwall tends to force air in the upwind end and out the downwind end, creating drafts along the building length, Fig 68. In MOF buildings, make every fifth pen cross partition solid (about every 50′). In open-front buildings, make every third cross partition solid (about every 25′). Extend the solid partitions up to the bottom of the sidewall vent doors and make them completely solid, even over the slats. If drafts are caused by wind swirling around the ends or over the top of an open-front building, close up 2′-3′ of the top and 16′ of each open wall end.

Do not block summer cross ventilation—make alley partitions open or "porous". Open mesh partitions below the front vent doors of a MOF building are ineffective—if used, cover them in winter.

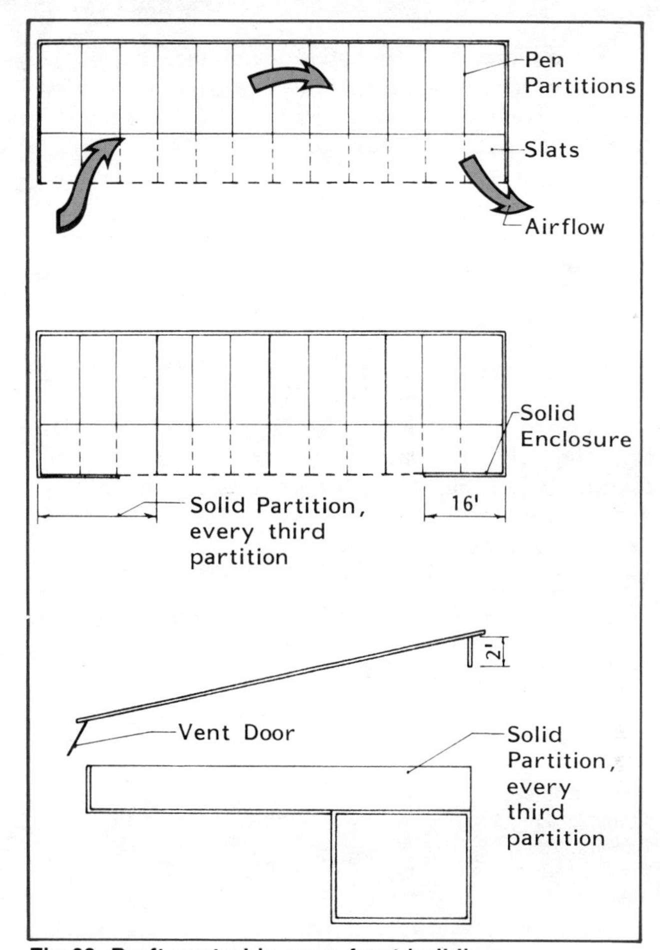

Fig 68. Draft control in open-front buildings.
Solid partitions and corner and top extensions reduce drafts.

MANURE MANAGEMENT

A complete manure disposal system:
- Maintains good animal health through sanitary facilities.
- Avoids air and water pollution.
- Complies with local, state, and federal regulations.
- Balances capital investment, labor, and nutrient use.

Manure Production

To estimate total manure production in a farrow-to-finish operation, multiply the number of productive sows by 105 to get lb manure produced/day; multiply the number of productive sows by 1.75 to find ft^3 manure/day. Table 18 gives more exact manure production values.

Table 18. Manure production of swine.
Solids and liquids, including 15% extra from waterers and washwater. Average density of manure is 60 lb/ft^3.

18a. Individual animals.

Animal type	Average weight lb	Manure per head ft^3/day	ft^3/yr
Sow and litter	400	0.66	241
Prenursery pig	20	0.03	11
Nursery pig	55	0.07	26
Growing pig	115	0.14	52
Finishing pig	190	0.24	88
Gestating sow	325	0.20	73
Boar	400	0.25	91

18b. Swine operations.
Based on 16 pigs sold/productive sow/yr. Manure production per productive sow accounts for all animals in the operation such as boars and her pigs in nursery, growing, etc.
1 ft^3 = 7.5 gallons.

Operation type	Manure ft^3/yr	Total nitrogen lb N/yr
Feeder pigs produced, sold at 50 lb (per productive sow)	170	70
Pigs fed 50 to 220 lb (per pig sold/yr)	30	12
Farrow to finish (per productive sow)	650	266
(per pig sold/yr)	40	16

Solid Manure Handling

Solid manure results from catching and holding manure in bedding, or by allowing liquids to run off, leaving solids to be handled separately. Handling solid manure requires less equipment but more labor than liquid systems. Suggestions:
- Haul manure to fields whenever possible, but avoid spreading on frozen slopes near ditches, streams, and roads.
- When a manure stockpile is necessary, locate it out of natural drainageways, away from water sources, and for convenient loading into a spreader. Divert surface runoff around storage area.
- Control runoff from manure stockpiles or feedlots.

Slurry Manure Handling

Unbedded solid manure systems result in a slurry with a consistency between a solid and liquid. Slurries are difficult to store and handle, but can be picked up with a bucket front-end loader and spread with a flail-type spreader.

Avoid slurry manure systems. Use bedding and slope floor to drain off liquids if a solid system is desired. Scrape manure to a pit for dilution if a liquid system is desired.

Liquid Manure Handling

Liquid manure advantages:
- Requires less time and labor.
- Manure disposal can be postponed to fit field schedules, soil conditions, and expected rainfall, **if** the storage unit is properly sized.
- Odors, unsightliness, and flies can be controlled with covered storage and soil injection land application.

Flushing Systems

In a flush system, a large volume of water flows from one end of a building to the other, down a sloped, shallow gutter. The water scours manure from the gutter and removes it to a lagoon, Fig 69. There are two types:
- Open gutter, used primarily in finishing facilities.
- Underslat gutter, used in farrowing, nursery, gestation, and larger facilities where residue or disease transmission is a concern.

Gutter slopes range from flat to over 2%, but must either have no cross slope or a very slight crown. Limit gutter lengths to less than 150′. Gutter widths vary, depending on the application. Provide at least 1 ft^2 of gutter area per pig in open gutters. Subdivide wide underslat gutters (4′ and wider) into 2′-2½′ wide channels to ensure uniform cleaning, Fig 70. Use inverted concrete slats as channel dividers.

Flush with at least 30 gal/ft of gutter width for open gutters and at least 50 gal/ft of gutter width for underslat gutters. Flush tanks should release the entire water volume in 10 to 20 seconds.

Select total daily flush volume from Table 19. Divide the total daily volume by the volume per flush to determine the number of flushes per day.

For large capacity pump flushing, design at 80 gpm per foot of gutter width. Flush for 2 to 3 minutes at least twice a day. Pump flushing uses much more water than tank flushing.

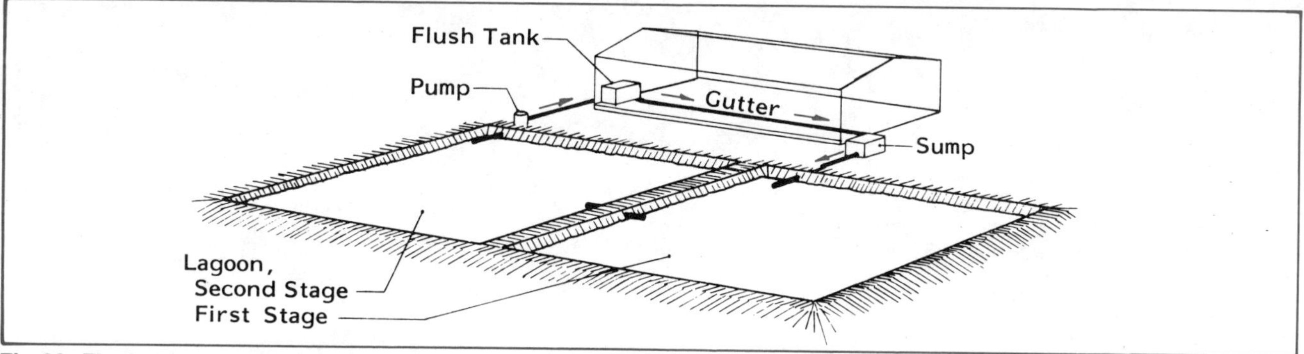

Fig 69. Flush system with two-stage lagoon.

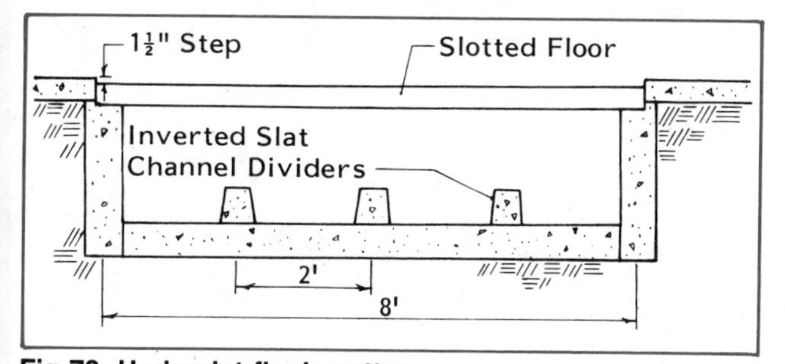

Fig 70. Underslat flush gutter.
Use channel dividers for all but the first 20′ of gutter.

Hang weighted flexible curtains in the gutter directly under partitions and outside building walls to reduce airflow through the gutter.

Mechanical Scrapers

Barn cleaners and alley scrapers can also remove manure from open or underslat gutters. They have fewer odors than pits because of frequent manure removal (at least once a day), but some ammonia is still released from the wet gutter surface. The mechanical equipment requires more maintenance than other manure handling systems. Outside manure storage is required.

Mechanical scrapers are well suited to remodeling, because their gutters are flat and have a minimum depth of 4″ in open gutters and 12″-24″ in underslat gutters. Removable slat sections facilitate repair. Hang weighted flexible curtains in the gutter under all building walls to reduce airflow through the gutter.

Narrow Gutters

There are two types of narrow gutters—deep narrow gutters and gravity drain gutters.

Deep narrow gutters are located along solid feeding floors of growing-finishing buildings, Fig 71. Wash or scrape manure into gutters. Put waterers near the gutter and slope the floor toward the gutter at 1″/ft. Slope the upper portion of the floor at ¼″/ft if floor feeding. Limit pen length to 20′ to reduce daily scraping.

Gravity drain gutters are under slotted floors, mainly in farrowing and nursery buildings, Fig 72. Gravity drain gutters have Y, U, V, or rectangular shapes. Paint the sloping sides of concrete gutters to make them slick for better cleaning.

Table 19. Total daily flush volume.

Swine type	Flush volume gal/hd
Sow and litter	35
Prenursery pig	2
Nursery pig	4
Growing pig	10
Finishing pig	15
Gestating sow	25

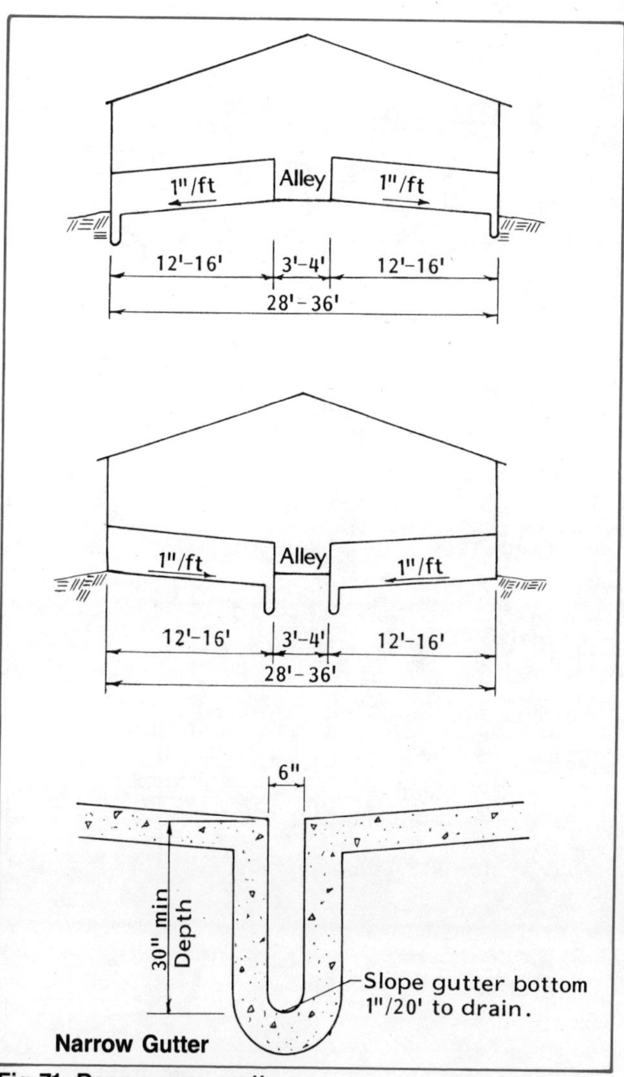

Fig 71. Deep narrow gutters.
Gutters are self-cleaning if flushed when full.

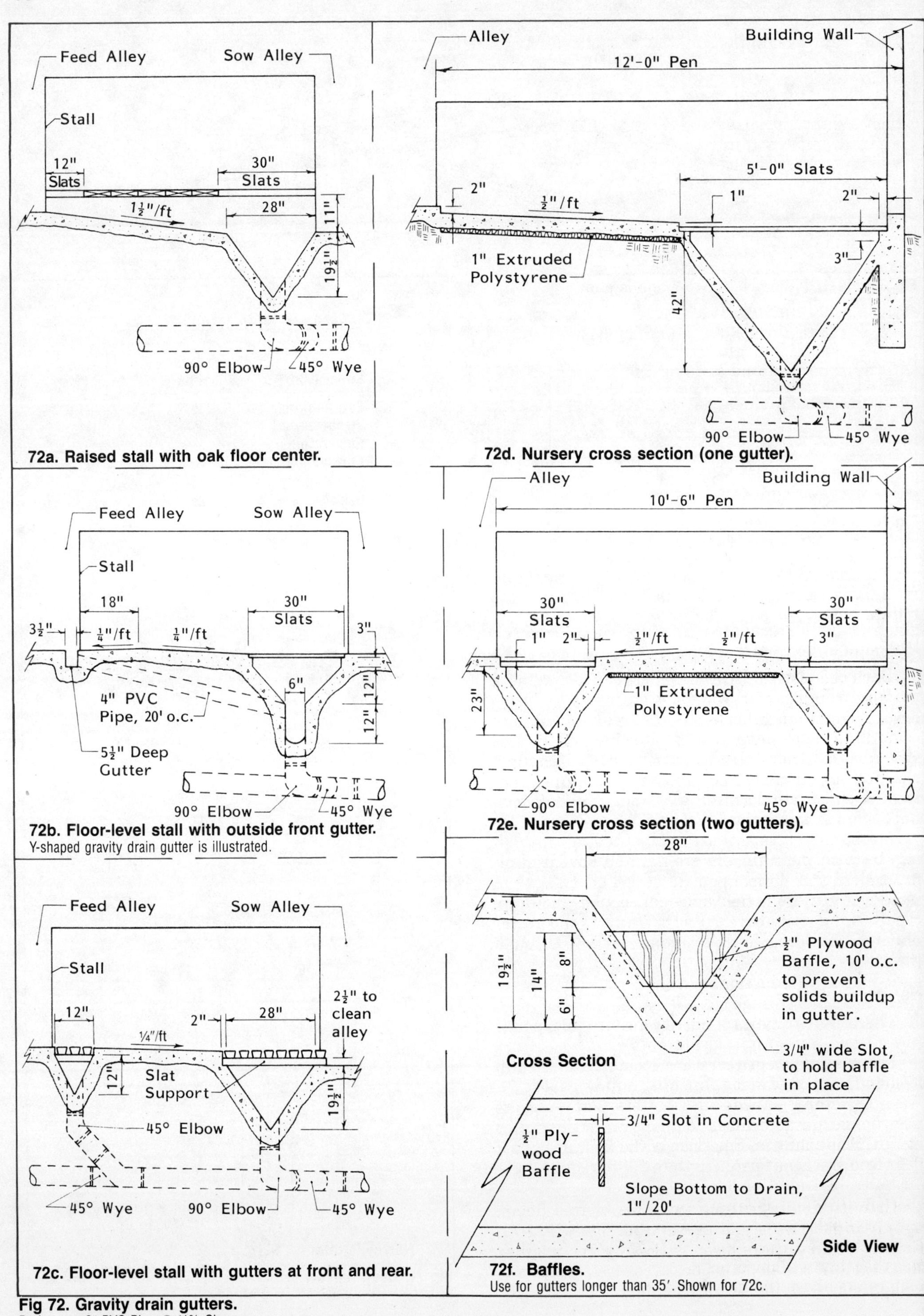

Fig 72. Gravity drain gutters.
Drains are: 8" PVC Pipe @ ½% Slope or
6" PVC Pipe @ 1% Slope

Limit gutter lengths to about 50′ and slope bottom of gutter towards the drain at 1″/20′. When gutter is full (usually 3 days to 1 week), pull the drain plug—manure flows by gravity to an outside storage tank or lagoon. Agitation is usually not needed before draining narrow gutters. Minimum drain pipe diameter is 6″, with 8″ preferred. Slope 6″ diameter pipes at 1%; 8″ at ½%.

Trickle Overflow

A trickle overflow system is suitable for farrowing and nursery buildings. In theory, biological action liquefies settled solids and keeps them moving toward a standpipe overflow. Two storages are required—a 2′-3′ deep underslat pit and an outside tank or lagoon. Biological activity does not always perform as desired and solids may build up. Install manure ports 20′ o.c. to operate as a liquid manure pit if necessary.

Recirculation

Recirculation systems are usually converted trickle overflow units that have malfunctioned. Set the pit overflow to maintain about 1′ of liquid depth. Drain the pit to a lagoon every 3 to 7 days and refill immediately with lagoon water. Partition the pit into sections less than 100′ long; fill and drain each section individually.

Liquid Manure Storages

Liquid manure storages usually are underslat pits or outside tanks. Outside liquid manure storage may be earth or prefabricated concrete or steel. Prefabricated storage generally costs more per unit capacity than a pit under slats because of the manure transfer equipment. Uncovered earth pits have a large surface area and often produce a lot of odor. Elevated storage tanks disperse odors better. Lagoons are discussed in another section.

The storage capacity required depends on number and size of pigs, amount of spilled and cleaning water, and desired storage period. Plan for 4 to 6 month's capacity if manure is field spread, to avoid spreading on frozen or snow-covered ground or on growing crops.

Storage capacity equals (number of animals) x (daily manure production) x (desired storage time in days). Add 12″ to storage tank depth as a safety factor and to account for manure that is left after pumping.

For buildings with pit ventilation, add another 12″ to pit depth. Pit depth is the distance between pit floor and bottom of slat support beams.

For uncovered outside storage tanks, increase capacity to account for rain and extra dilution if irrigating with the manure. See section on lagoons.

Locate storages for convenient filling, emptying, and controlled addition of dilution water. Locate storages downhill and as far as practical from the water supply (check local regulations for minimum distances). Avoid creviced limestone, shale, and bedrock sites that might allow direct ground water pollution. Keep tanks above the normal water table and out of

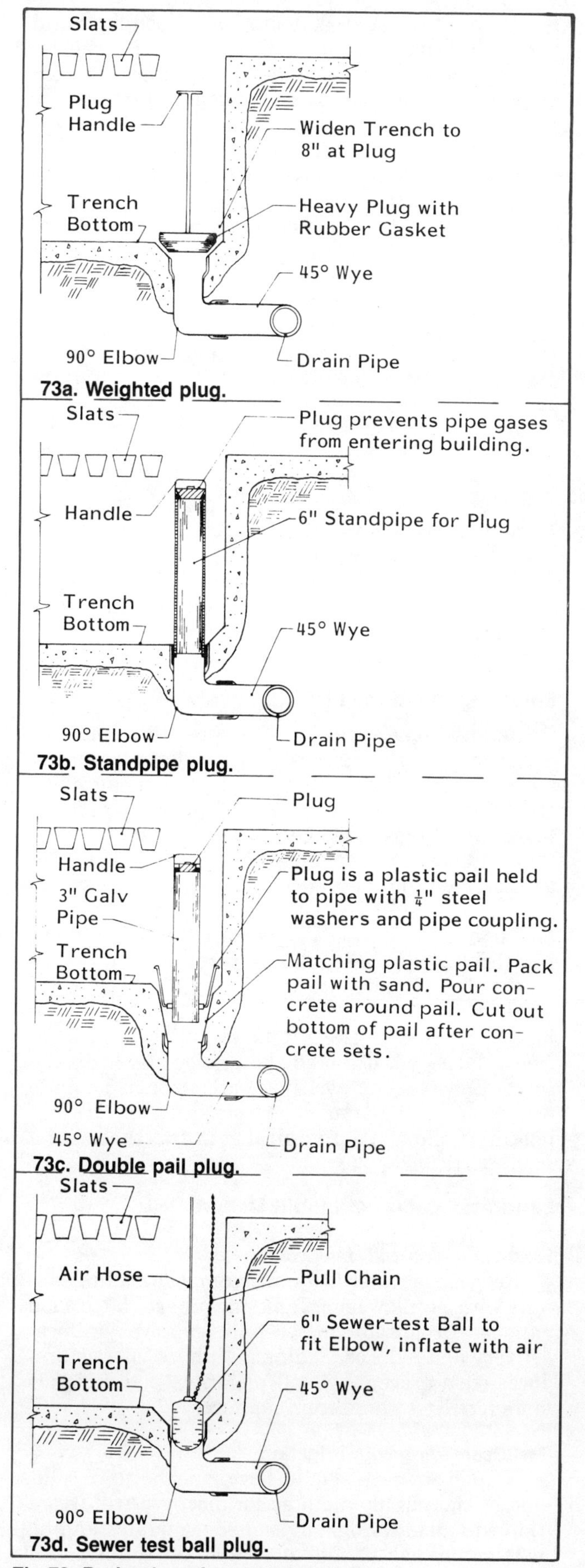

Fig 73. Drain plugs for narrow gutters.

flood plains to avoid tank floatation or flooding. Build part of the tank above ground and backfill to keep the pit floor above the water table. Obtain approval from appropriate regulatory agencies before starting construction.

Avoid any slope or depressed pump-out area in the manure tank floor to prevent a buildup of solids there. Place vacuum tank manure cleanout ports 20′ o.c. Place agitator/chopper pump outlets no more than 75′ apart around the pit perimeter. Protect necessary tank openings with airtight covers to keep out children, animals, equipment, and other objects and to prevent short circuiting of ventilating air. For construction details on rectangular reinforced concrete manure storages, get mwps-74303, *Liquid Manure Tanks*.

When construction is complete, clean out nails and other foreign material that could damage pumps. Add water before filling with manure: at least 3″-4″ to pits under slotted floors and 6″-12″ if loaded with batches of scraped wastes. Keep solids submerged to discourage fly reproduction.

Scrape manure into storage frequently to keep it wet. Dry manure is difficult to reliquefy. Do not load frozen manure, fibrous material, or debris that may interfere with agitation or pumping.

Emptying liquid manure storages

Because some solids settle, agitate stored manure just before emptying. Most solids buildup problems occur in growing, finishing, and gestation buildings. **CAUTION: Agitation of stored manure can release deadly gas. See section on safety.**

Agitator/chopper pumps operating at about 2000 gpm provide the most effective agitation, but pump annexes are required. Some agitator pumps operate at 900 to 1800 gpm and can operate through conventional 8″-12″ diameter manure ports.

Divide large pits into chambers no more than 40′ on a side to improve agitation. Locate overflows at the top of partitions between pits, so pumping an adjacent pit does not drain off the liquids and leave the solids. Construct pit partition walls strong enough to withstand a full manure load from either side. Do not put an overflow in a pit wall that is directly under a room partition.

Land Application of Liquid Manure

Field spreading with tank wagon

Keep extra water to a minimum if spreading manure with a tank wagon. Tank wagons are 750 to 4500 gallons. An agitator in the tank improves uniform delivery and reduces plugging. Manure injection reduces odor, prevents runoff pollution, and reclaims more fertilizer value from manure.

Field spreading with irrigation

Liquid manure should have less than 5% solids and be thoroughly agitated for most types of irrigation equipment. Some helical screw pumps and big nozzle guns can handle more than 5% solids. Select spreading sites with caution, because odors can be severe. Overapplication of manure can result in runoff and groundwater pollution.

Lagoons

Lagoons treat as well as store manure before land application. Use two-stage lagoons for recycled flush water to improve flush water quality and decrease levels of pathogenic organisms. Dilution water improves treatment and odor control—add as much dilution water as manure. Empty at least ⅓ of lagoon volume once or twice a year. Properly designed and managed lagoons have a useful life of 8 to 12 years before filling with solids. See Fig 74 and Tables 20-22 for design information.

Example: Design a two-stage lagoon for a central Iowa farrow-to-finish operation which markets 2000 head per year.

1. From Fig 74, central Iowa is in Zone 2.
2. From Tables 20 and 21, size the first stage lagoon at 210 ft^3/head x 2000 head = 420,000 ft^3 and the second stage at 60 ft^3/head x 2000 head = 120,000 ft^3.
3. First stage: 420,000 ft^3 volume. Entering Table 22 on the 425,000 ft^3 row and for a 15′ lagoon depth, select 200′ x 236′ x 15′.
4. Second stage: 120,000 ft^3 volume. Entering Table 22 on the 120,000 ft^3 row, select 150′ x 143′ x 10′.
5. Increase total dimensions to include berm widths and exterior berm sideslopes.

Table 20. Sizing single stage lagoons.
Also use for first stage of two-stage lagoons. Always provide 1′-2′ freeboard.

20a. Individual animals.

Animal type	Weight lb	Climatic zone (Fig 74) 1	2	3	4	5	6	7
		- - - - - - - ft^3/hd- - - - - -						
Sow and litter	400	1470	1280	1130	1024	940	855	790
Prenursery pig	20	55	48	42	39	35	32	30
Nursery pig	55	150	132	116	105	96	88	82
Growing pig	115	320	275	245	215	200	184	170
Finishing pig	190	530	455	405	355	330	305	280
Gestating sow	325	450	405	355	310	285	260	235
Boar	400	550	480	435	390	340	320	300

20b. Swine operations.
Based on 16 pigs sold/productive sow-yr. Manure production per productive sow accounts for all animals in the operation such as boars and her pigs in nursery, growing, etc.

Operation type	Climatic zone (Fig 74) 1	2	3	4	5	6	7
Feeder pigs produced, sold at 50 lb							
(ft^3/productive sow)	1050	950	850	750	700	600	550
Pigs fed 50 to 220 lb							
(ft^3/pig sold/yr)	175	150	135	115	110	100	95
Farrow to finish							
(ft^3/productive sow)	3850	3350	3000	2600	2400	2200	2050
(ft^3/pig sold/yr)	240	210	185	165	150	140	130

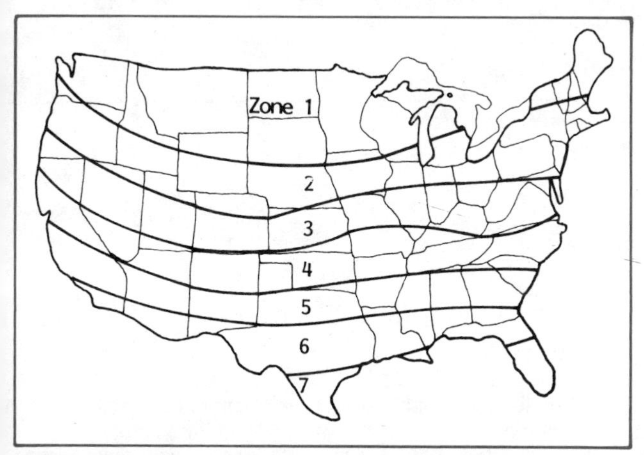

Fig 74. Climatic zones for anaerobic lagoons.
Based on volatile solids loading rates recommended by the American Society of Agricultural Engineers.

Table 21. Sizing two-stage lagoons.
Use Table 20 to size the first stage of a two-stage lagoon. This table is for typical Midwest climate where evaporation is about equal to rainfall. Provide an extra 1′-2′ of depth if rainfall greatly exceeds evaporation in your area. Always provide 1′-2′ of freeboard.

21a. Individual animals.

Animal type	Weight lb	ft^3/hd
Sow and litter	400	345
Prenursery pig	20	16
Nursery pig	55	36
Growing pig	115	75
Finishing pig	190	125
Gestating sow	325	105
Boar	400	130

21b. Swine operations.
Based on 16 pigs sold/productive sow-yr. Manure production per productive sow accounts for all animals in the operation such as boars and her pigs in nursery, growing, etc.

Operation type	ft^3 per pig sold-yr	ft^3 per productive sow
Feeder pigs produced, sold at 50 lb	—	300
Pigs fed 50 to 220 lb	45	—
Farrow to finish	60	1000

Runoff Control For Open Lots

Runoff may cause pollution and is regulated. Follow these rules if you have no specific regulations in your area.

- Consider residences and zoning before building.
- Locate feedlots away from streams—at least far enough for construction and operation of adequate detention structures and to permit future expansion.
- Locate the lot at or near the top of a slope or divert upslope water to prevent outside drainage from crossing the lot.

Control lot runoff in two stages—one for solids and another for liquids. A settling (or debris) basin removes solids. Liquids go to a holding pond for storage until field spreading or to a vegetative infiltration area where they infiltrate into the soil.

Settling Basins

Settling basins allow larger solids to settle and liquids to drain. They remove 50%-85% of the solids from lot runoff, so fewer solids settle in the holding pond or vegetative infiltration area. Solids removal reduces odors from holding ponds and makes liquids easier to pump with smaller irrigation equipment.

The best settling basin is relatively large and shallow. A concrete bottom or a complete concrete basin is necessary if equipment enters the basin for solids cleanout. Slope basin access ramps no more than 1″ fall/1′ run for front-end loaders. Recommended settling basin capacity is 5 to 10 ft^3 / 100 ft^2 of feedlot area. Design for at least 30 minutes of detention to settle most runoff solids.

If liquid manure equipment is available on the farm, consider handling the semi-solids in the settling basin as liquid manure. Periodically, agitate and pump out the basin. Cleaning frequency depends on basin size and type, lot slope, amount of lot manure, and storm runoff characteristics.

In dewatering basins, perforated or slotted openings drain liquids to allow solids to dry, Fig 75. Remove solids after each major storm. Another alternative is a weir notch in the basin sidewall, Fig 76. With the weir notch overflow, liquid remains in the basin after a storm. Clean with liquid manure handling equipment 3 or 4 times a year.

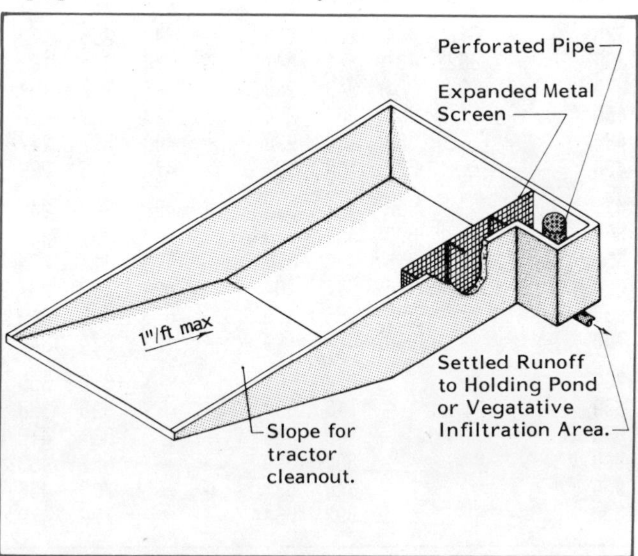

Fig 75. Concrete settling basin.
Dewatering type—solids are removed with solid manure handling equipment.

Table 22. Top dimensions for lagoons.
Volume based on 1′ of freeboard and 2.5:1 slope. All lagoon dimensions in feet.

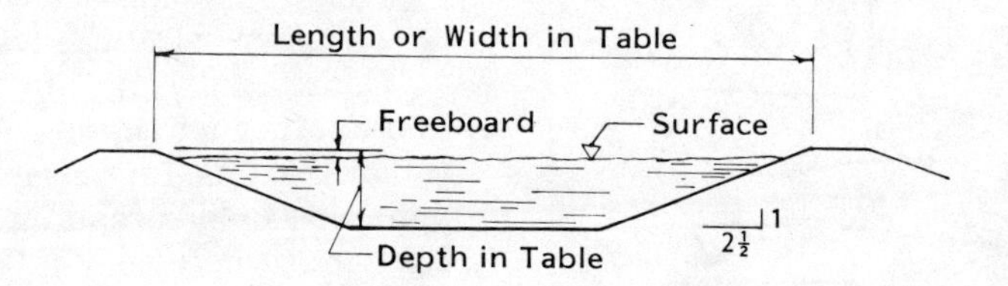

Interior length of lagoon when—

Lagoon volume Thousands of ft³	Depth is 6′ and Interior width is					Depth is 10′ and Interior width is					Depth is 15′ and Interior width is					Depth is 20′ and Interior width is				
	100	150	200	300	400	100	150	200	300	400	100	150	200	300	400	100	150	200	300	400
10	44	33	-	-	-	-	-	-	-	-	-	-	-	-	-	-	-	-	-	-
20	73	51	41	32	-	56	-	-	-	-	-	-	-	-	-	-	-	-	-	-
30	102	70	55	41	34	73	53	-	-	-	-	-	-	-	-	-	-	-	-	-
40	132	88	68	49	40	89	63	52	-	-	81	-	-	-	-	-	-	-	-	-
50	161	107	82	58	47	106	73	59	-	-	93	-	-	-	-	-	-	-	-	-
60	191	125	95	67	53	123	83	67	51	-	105	75	-	-	-	103	-	-	-	-
70	220	144	109	76	60	139	93	74	56	-	118	82	-	-	-	114	-	-	-	-
80	249	162	122	85	66	156	103	81	60	51	130	89	-	-	-	125	-	-	-	-
90	279	181	136	93	73	173	113	88	65	54	142	95	77	-	-	136	-	-	-	-
100	308	199	149	102	79	189	123	95	69	57	154	102	82	-	-	147	-	-	-	-
110	338	218	163	111	86	206	133	102	74	61	167	109	87	-	-	158	104	-	-	-
120	367	236	176	120	92	223	143	109	79	64	179	116	92	-	-	169	109	-	-	-
130	396	255	190	128	99	239	153	117	83	67	191	123	96	-	-	180	115	-	-	-
140	426	274	204	137	105	256	163	124	88	71	204	130	101	77	-	192	121	-	-	-
150	455	292	217	146	112	273	173	131	92	74	216	136	106	80	-	203	126	101	-	-
160	-	311	231	155	118	289	183	138	97	77	228	143	111	83	-	214	132	104	-	-
170	-	329	244	164	125	306	193	145	101	81	241	150	115	85	-	225	137	108	-	-
180	-	348	258	172	131	323	203	152	106	84	253	157	120	88	-	236	143	112	-	-
190	-	366	271	181	138	339	213	159	110	87	265	164	125	91	76	247	148	115	-	-
200	-	385	285	190	144	356	223	167	115	91	278	171	130	94	78	258	154	119	-	-
225	-	431	318	212	161	398	248	184	126	99	308	188	141	102	84	286	168	128	-	-
250	-	477	352	234	177	439	273	202	138	107	339	205	153	109	89	314	182	138	102	-
275	-	524	386	256	193	-	298	220	149	116	370	222	165	116	94	342	196	147	108	-
300	-	570	420	278	209	-	323	238	160	124	401	239	177	124	100	369	209	156	113	-
325	-	616	454	299	225	-	348	256	172	132	431	256	189	131	105	397	223	165	119	-
350	-	662	487	321	242	-	373	274	183	141	462	273	201	138	110	425	237	175	125	103
375	-	709	521	343	258	-	398	292	194	149	-	290	212	146	116	453	251	184	130	107
400	-	-	555	365	274	-	423	309	206	157	-	307	224	153	121	-	265	193	136	111
425	-	-	589	387	290	-	448	327	217	166	-	324	236	160	126	-	279	202	141	115
450	-	-	622	409	307	-	473	345	229	174	-	342	248	168	132	-	293	212	147	119
475	-	-	656	431	323	-	498	363	240	182	-	359	260	175	137	-	307	221	152	123
500	-	-	690	453	339	-	523	381	251	191	-	376	272	182	142	-	321	230	158	127
550	-	-	758	497	372	-	573	417	274	207	-	410	295	197	153	-	348	249	169	135
600	-	-	825	541	404	-	623	452	297	224	-	444	319	211	163	-	376	267	180	143
650	-	-	893	585	436	-	673	488	319	241	-	478	343	226	174	-	404	286	191	151
700	-	-	960	628	469	-	723	524	342	257	-	513	366	241	185	-	432	304	202	159
750	-	-	-	672	501	-	-	559	365	274	-	547	390	255	195	-	459	323	213	167
800	-	-	-	716	534	-	-	595	388	291	-	581	414	270	206	-	487	341	225	175
850	-	-	-	760	566	-	-	631	410	307	-	615	437	285	216	-	515	360	236	182
900	-	-	-	804	599	-	-	667	433	324	-	649	461	299	227	-	543	378	247	190
950	-	-	-	848	631	-	-	702	456	341	-	683	485	314	238	-	571	397	258	198
1000	-	-	-	892	664	-	-	738	479	357	-	718	508	329	248	-	598	415	269	206

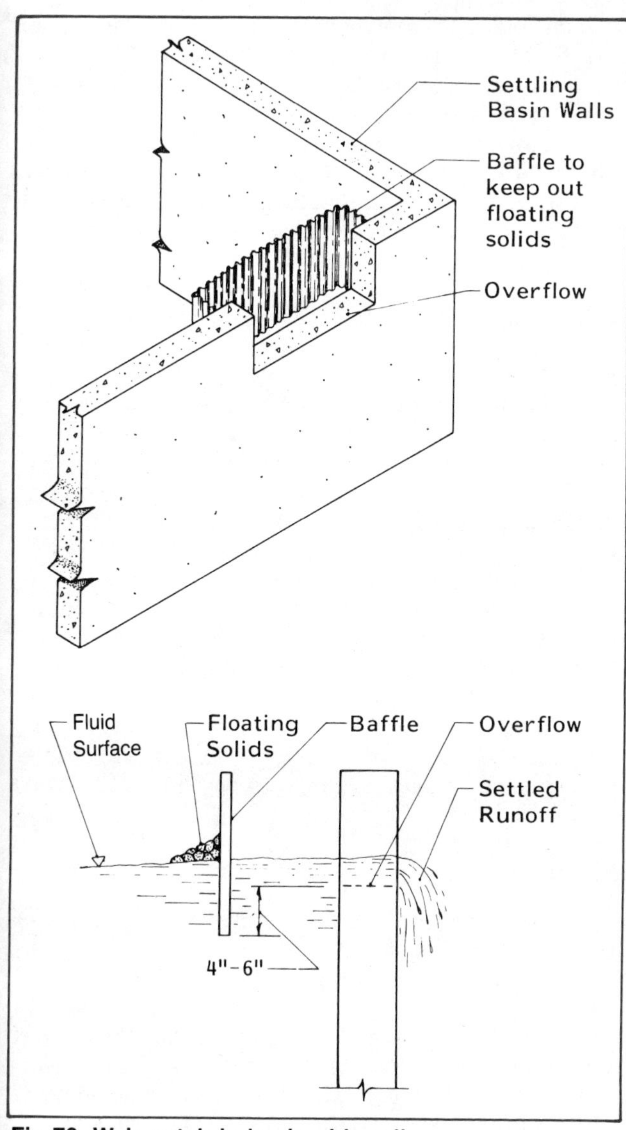

Fig 76. Weir notch in basin sidewall.
Agitate to suspend solids and clean with liquid manure handling equipment.

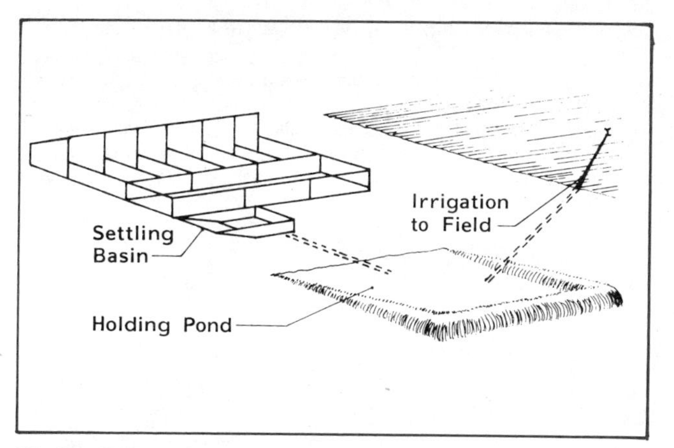

Fig 77. Holding pond.

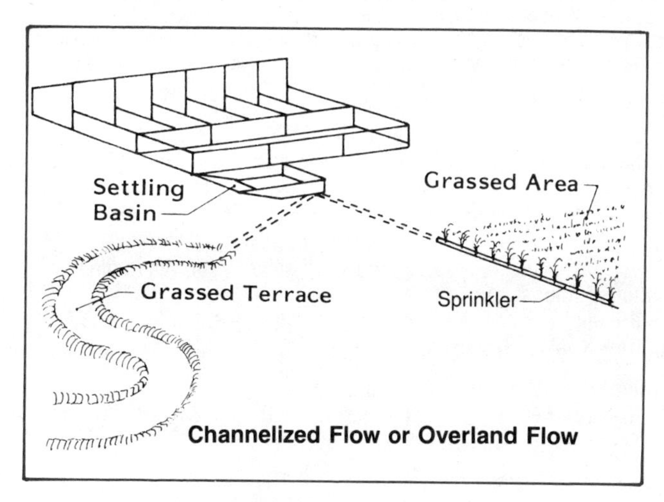

Fig 78. Vegetative infiltration area.

Holding Ponds

A holding pond temporarily stores runoff water from a settling basin, Fig 77. Design for at least the storage time required in your state (usually 90 to 180 days). With additional capacity, emptying the pond can be delayed to fit labor and cropping schedules without fear that another runoff event will cause overflow.

Holding ponds do not treat manure as in a lagoon, but simply store it until disposal. Empty a holding pond completely when pumping. Holding ponds tend to seal naturally, so seepage and ground water pollution are usually not a problem. If your pond is in sand or gravel soils or near fractured bedrock, seal the bottom by lining with plastic or 6" of compacted clay. Before construction, consult local Soil Conservation Service and regulatory agencies for assistance and information.

Vegetative Infiltration Areas

On many sites, an adequately sized grassed or cropped area downslope from the lot can dispose of runoff economically. A settling basin to remove solids is essential. Fig 78.

The infiltration area can be a long, 10'-20' wide channel or a broad, flat area. In either case, provide at least 300' of travel from the inlet of the infiltration area to a ditch or stream. Slope the first 50' at about 2% (1' in 50') to prevent settling of solids near the settling basin outlet. Slope the rest of the channel at about 0.25% to prevent standing water.

Land Application of Manure

Elemental nutrients per 1000 gal of raw manure are about 55 lb total nitrogen (N), 18 lb phosphorous (P), and 32 lb potassium (K). The fertilizer value of manure when spread on land is less, depending on how it is collected, stored, and spread. See Table 23. Avoid spreading manure on frozen ground where runoff from rain or spring thaws could pollute a waterway.

Table 23. Nutrient value of liquid swine manure, approx.
Nutrient values vary considerably with dry matter content. These values include substantial amounts of dilution water. Nutrient losses from storage and handling are included, i.e. values are at time of land application. Available nitrogen is primarily ammonium-N and 35% organic nitrogen which is available to plants during the growing season. Multiply P_2O_5 by 0.44 to convert to elemental P; multiply K_2O by 0.83 to get elemental K. From Pork Industry Handbook (PIH-25), "Fertilizer Value of Swine Manure."

Manure from	Dry Matter %	Available N	Total N	P_2O_5	K_2O
		------lb/1000 ft^3------			
Liquid pit	4	195	270	203	165
Lagoon	1	30	30	15	30

Table 24. Land application of manure.
Based on 200 lb of usable N/acre-yr (50% of applied total N assumed to be usable). Manure nitrogen levels account for storage and application losses. Assumes 16 pigs sold/productive sow-yr. Manure production per productive sow accounts for all animals in the operation such as boars and her pigs in nursery, growing, etc. Based on PIH-25, "Fertilizer Value of Swine Manure."

Manure handling method	Feeder pigs produced, sold at 50 lb	Farrow to finish	Pigs fed 50 to 220 lb	Farrow to finish
	Acres per 10 productive sows		Acres per 100 pigs sold/yr	
Slotted floor & pit:				
Broadcast	3.1	4.8	2.9	3.8
Knifed or cultivated	4.0	6.3	3.8	5.0
Solid floor & outside lot (Scraped Manure):				
Broadcast	1.9	2.9	1.8	2.3
Broadcast & cultivated	2.3	3.4	2.2	2.8
Lagoon (irrigated)	0.8	1.2	0.7	1.0

Controlling Odor

Locate swine buildings and manure storage where prevailing summer winds and air drainage downhill on calm nights will not carry odors to you or your neighbor. In most of the Midwest, locate manure storages to the north or east of residences. There is no "safe" separation distance, but try to locate your facilities at least ½ mile from neighbors—farther if you have a large operation, open lot, or lagoon.

Open lots have more odor than environmentally controlled buildings, especially after rain. Frequent scraping reduces odors because piled manure has less surface area for odor production.

Control ventilation exhaust odors by locating fans high on walls or placing buildings on high ground to disperse odors. Frequent removal of manure and proper pit ventilation reduces odors within the building.

Lagoon odors can be reduced with proper dilution and loading but can still be severe during spring and fall when lagoon contents change temperatures and turn over.

Use soil injection to reduce land application odors. Spray irrigation produces high odor levels. Spread or irrigate only on days when a brisk wind is blowing away from residential areas. Spread in the morning, because of faster drying and more favorable air currents.

Shield swine buildings and lagoons from public view and practice "good neighbor" policies. Christmas hams and plowing out snowbound neighbors may help reduce confrontations. Use good housekeeping on the farm to present a good image.

To date, experience shows most "odor control" products for manure pits are of little value. If you try one of these products, arrange to pay only after you are satisfied with the results.

Safety

Gases from liquid manure storage units may be hazardous. The most serious problems occur when manure is agitated or when ventilating fans fail.

Provide maximum ventilation when agitating or pumping manure. If possible, pump when no animals are in the building. Agitate and pump on milder days during cold seasons to reduce heat lost with the higher ventilating rate.

Enter a storage tank only if absolutely necessary and only with adequate safety precautions. Wear self-contained breathing equipment such as scuba diving gear. Chemical reaction filter masks are not sufficient. Before entering, ventilate the pit thoroughly. Always wear a safety line and work with someone strong enough to hoist you out of the pit if you have trouble.

SLOTTED FLOORS

Slat Width and Spacing

Recommended slat width depends on slat material and pig size. The larger the pig, the wider the slats can be without sacrificing performance. Some metal and fiberglass slats for small pigs are as narrow as 1″.

Use 4″ or narrower slats for sows in stalls to ensure adequate drying and cleaning. Provide a very slight crown on concrete slats to improve cleaning. See Table 25 for concrete slat and slot widths.

Table 25. Recommended slat and slot widths.
Avoid slots between ⅜″ and ¾″ wide. Wire mesh, metal, or plastic slats are preferred for farrowing and prenursery.

Animal type	Slot width in.	Concrete slat width, in.
Sow and litter	⅜	4
Prenursery pig	⅜	Not recommended
Nursery pig	1	4
Growing-finishing pig	1	6-8
Gestating sow or boar:		
Pens	1	6-8
Stalls	1	4

Slat Direction

In partly slotted buildings, place slats parallel to the long dimension of rectangular pens. In farrowing and gestation stalls, place slats parallel to the sow.

Slat direction in totally slotted buildings are influenced by other factors, such as construction costs. However, try to place slats parallel to natural traffic flow, so pigs walk along rather than across slats.

Slotted Floor Materials

Wood slats have the lowest initial cost and the shortest life. Oak is best, but other hardwoods such as hickory and maple may be used. Pig performance may be reduced by variable spacing due to warping (particularly with slats less than 3″ wide), wear, careless installation, insecure fastening, and uneven slat heights. Wood slats may be slick. Their life expectancy is 2 to 4 years and even less in areas of intensive use, such as around feeders and waterers. Recommended dimensions are in Table 26.

Table 26. Native oak slat dimensions.
Design load: 100 lb/linear ft; swine up to 200 lb.

Slat dimensions laid flat BxA, in.	Maximum span ft.
1½″x3½″	3
1½″x5½″	4

Concrete slats are the most durable and work especially well for hogs over 75 lb, including gestating sows, Table 27. Precast slats may have higher quality material and construction. Some companies form gang slats (two or more slats connected into one section) to reduce installation labor.

Table 27. Concrete slat design.
Design load: 125 psf; includes swine up to 200 lb and slat dead load. Bottom width is 1″ less than top width. Minimum concrete compressive strength is 3500 psi; minimum steel yield point is 40,000 psi.

Span ft	Slat dimensions Top width in.	Depth in.	Reinforcing bar
4′	4″	4″	No. 3(⅜″)
	6	4	No. 3
	8	4	No. 3
6′	4″	4″	No. 4(½″)
	6	4	No. 4
	8	4	No. 4
8′	4″	5″	No. 5(⅝″)
	6	5	No. 4
	8	5	No. 4
10′	4″	6″	No. 5
	6	6	No. 5
	8	6	No. 5

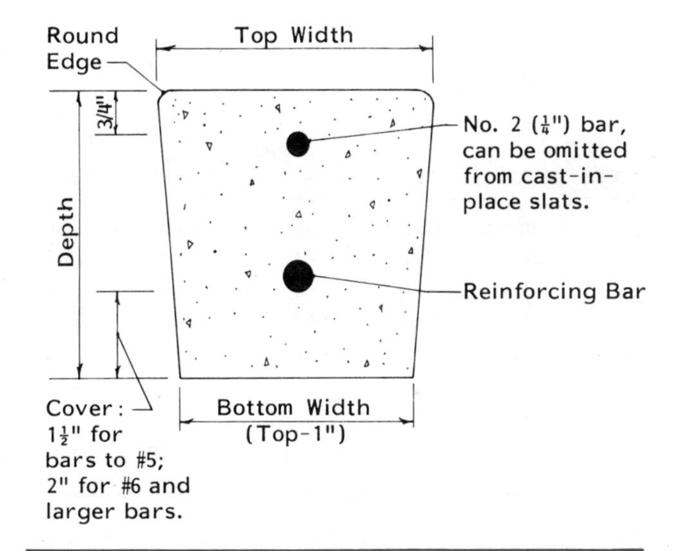

Use concrete with at least a 7½ bag mix of air-entrained cement, 2″-3″ slump, and ½″ maximum aggregate. Steel reinforcing is required, Table 27. Edging slats with a ¼″ sidewalk edger prevents chipping, improves cleaning, and reduces foot injuries. Give slats a smooth, steel trowel finish to reduce leg and knee abrasions. Cure slats for 5 days by covering with plastic, a curing compound, or wet straw.

Steel slats are adequate in certain applications. Many early steel slats failed after 2 to 4 years because the steel corroded from the underside. Epoxy paint on the underside of steel slats helps. Porcelain-coated steel slats resist corrosion, providing they are not chipped. Plastic-coated slats work well in farrowing stall creeps but some are short-lived under the sow.

Woven wire and steel rod mesh work well for farrowing and nursery facilities. Use an overlay in the pig creep the first few days after farrowing if foot injuries are a problem.

Flattened expanded metal (¾", 9 to 11 gauge) works for pigs under 30 lb, but does not last as long as concrete when subjected to concentrated traffic by heavier pigs. Avoid expanded metal with sharp edges. Flattened expanded metal is not recommended for sows with litters less than 14 days old because of foot, leg, and teat injuries.

Stainless steel also works well in farrowing and nursery units. Most stainless slotted planks are 8"-12" wide and span 4'-10' depending on flange depth. Stainless steel lasts considerably longer than plain or galvanized steel.

Aluminum slotted flooring is available in several shapes, from narrow "T's" to 8" wide punched planks. Aluminum cleans easily, but may be slick for heavier swine. Insulate or isolate aluminum from other materials, particularly steel, to avoid corrosion.

Plastic slotted flooring tends to be slick unless coated with a rough surface. Fiberglass reinforced plastic slats are popular for farrowing and nursery. The fiber strands interrupt the fatigue factor found in some plastics, making them more durable.

Floor Smoothness

Rough concrete floors result in more injury, lameness, and infection than smooth, clean floors. However, a floor can be too smooth. If the floor is kept dry, and the slope does not exceed ½"/ft, slickness problems are rare. An exception is breeding areas where footing is more critical. Refer to *Slip Resistant Concrete Floors,* AED-19, by Midwest Plan Service, for more information.

Latex, epoxy, or chlorinated rubber paints on the creep floor of farrowing stalls may reduce knee abrasions. Sprinkle sawdust on wet paint for additional traction. Check paint labels for toxic materials, such as lead. Repaint annually and avoid washing painted area with high pressure washers.

Floor Supports

Slotted floor beam and column support designs are shown in Tables 28 and 29. Table 30 gives designs for grid systems to support metal or plastic floors—use only if product literature is not available.

Table 28. Concrete beams for concrete slat floors.
Design load: 125 psf; includes swine up to 200 lb and slat dead load. Beams cast in place. Minimum concrete compressive strength is 3500 psi; minimum steel yield point is 40,000 psi.

Slat length between joists		Beam span 6'	8'	10'	12'
6'	WxD	6"x6"	6"x9"	8"x9"	8"x10"
	Bars	2#5	2#5	2#7	2#8"
8'	WxD	6"x8"	8"x8"	8"x9"	9"x10"
	Bars	2#5	2#7	2#8	3#8
10'	WXD	6"x9"	8"x10"	8"x12"	9"x11"
	Bars	2#5	2#7	2#8	3#8

W = width, in.
D = depth, in.
Bars = number of bars and size of bars in bottom of beam.
Cover = 1½" for sizes through #5;2" for #6 and larger.

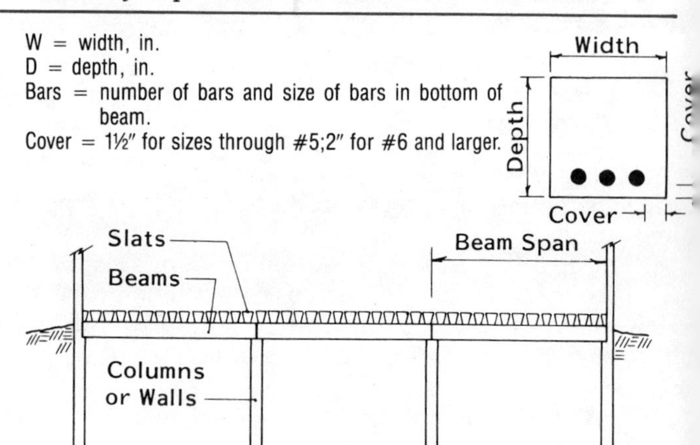

Table 29. Unreinforced concrete footings and masonry columns.
Design load: 125 psf; includes swine up to 200 lb and slat dead load. Block cores are filled. All footings are square. Soil bearing pressure = 3000 psf.

Slat length between beams	Column height	Beam span 6'	8'	10'	12'
		Column size			
6'	4'	8"x8"	8"x8"	8"x8"	8"x8"
	8'	8"x8"	8"x8"	8"x8"	8"x16"
8'	4'	8"x8"	8"x8"	8"x8"	8"x8"
	8'	8"x8"	8"x8"	8"x16"	8"x16"
10'	4'	8"x8"	8"x16"	8"x16"	8"x16"
	8'	8"x8"	8"x16"	8"x16"	8"x16"
		Footing, width x depth			
6'		18"x6"	20"x6"	22"x6"	24"x6"
8'		20"x6"	22"x6"	24"x6"	28"x8"
10'		22"x6"	24"x6"	28"x8"	30"x9"

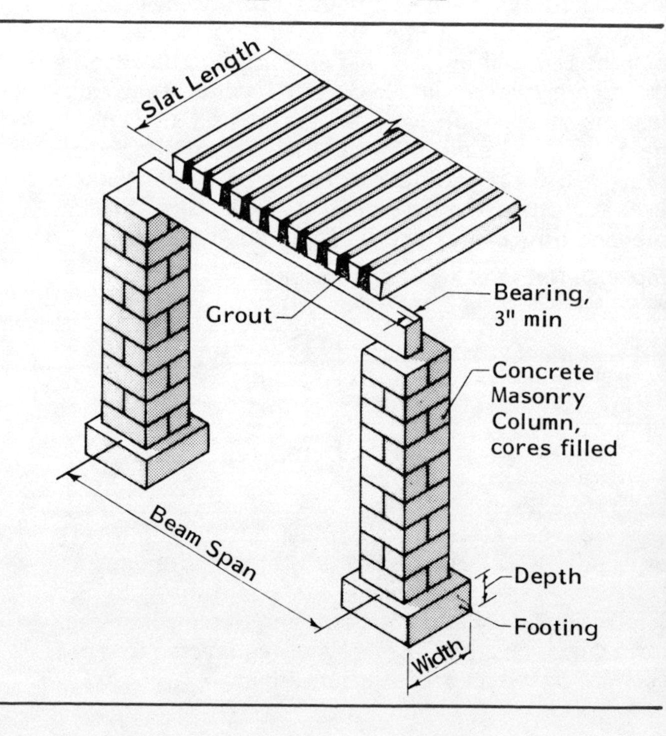

Table 30. Wood beams for wood, metal, or plastic slatted floors.

Design load: 60 psf; includes swine up to 75 lb and dead load. Bending strength is 1200 psi for 2x and 800 psi for 4x and 6x lumber. Mesh materials require blocking between the beams 12″ o.c. in farrowing and 16″ o.c. in a nursery. All beams set on edge.

Beam span	Beam size				
	2x4	2x6	2x8	2x10	2x12
	-------Max. beam spacing, ft.-------				
4′	2.55	6.3			
6′	1.13	2.8	4.87	7.92	11.7
8′	0.64	1.57	2.74	4.46	6.59
10′	0.41	1.01	1.75	2.85	4.22
	4x4	4x6	6x6	6x8	4x12
4′	3.97	9.8			
6′		4.36	6.85		
8′		2.45	3.85	7.16	10.25
10′		1.57	2.46	4.58	6.56

UTILITIES

Utilities for swine buildings include electricity, fuel, and water. Contact your insurance company for approved construction practices before installing a gas or electric system. Contact public health officials for approved water systems.

Electrical

Electrical system failure can suffocate animals in environmentally controlled buildings. Protect electrical system from excessive moisture, corrosive gases, rodents, physical abuse, and animal contact. **Surface mount** all electrical cable and fixtures and select equipment designed for humid, corrosive environments.

Circuit Cables:

- Use Type UF (underground feeder) cable for livestock buildings.
- Fasten cable to walls and ceiling with plastic-coated straps or staples.
- Conduit (plastic or metal) is not permitted in swine buildings by many insurance companies.

Outlet and Light Circuits

Use at least 12 ga wire for 120 volt (V) circuits. Protect each 12 ga wire circuit with a 20 amp (A) circuit breaker. Place no more than 10 outlets or lights (150 watt each maximum) on a 20 A circuit.

Provide one duplex ceiling outlet for every two farrowing stalls and every two pens (15'-20' o.c. along alley). Use only nonmetallic boxes suitable for outdoor use.

- Use dust- and moisture-tight incandescent lights and mount them in nonmetallic boxes.
- Use dust- and moisture-resistant fluorescent lights enclosed in gasketed covers. Do not install conventional fluorescent lights where temperatures may be below freezing.
- Clean light bulbs once a month to maintain efficiency.

Provide at least two light circuits in each room. Over pen areas use two uniformly spaced rows of lights on separate switches, so two light intensity levels are possible.

Space incandescent lights at two times their mounting height over pens, and 1.5 times their mounting height along alleys. Table 31 shows light requirements for specific areas. For example, a 24' wide farrowing building could obtain the required light level of 15 foot candles with (0.6x24) = 14.4 watts (W) of fluorescent light per foot of building length or (2.4x24) = 57.6 W of incandescent light per foot of building length.

White walls and ceilings improve light levels and are especially useful in areas where animals are inspected frequently. Place ceiling-mounted lights at least 4' from ventilation inlets, so they don't deflect incoming air.

Light photoperiods play an important part in swine reproductive functions. Gilts reach puberty earlier if housed at 15 to 20 foot candles for 16 hours per day. Not enough is known to give recommendations on photoperiods for growing pigs.

Table 31. Light levels for swine housing.
Based on an 8' high ceiling. Add lights for specific tasks such as desk work.

Application	Illumination foot-candles	Fluorescent watts/ft²	Incandescent watts/ft²
Farrowing	15	0.6	2.4
Nursery	10	0.4	1.6
Growing-finishing	5	0.2	0.8
Gilt pool	15	0.6	2.4
Breeding-gestation	15	0.6	2.4
Feed storage and processing	10	0.4	1.6
Record keeping/office	70	2.8	11.2
Animal inspection/handling	20	0.8	3.2

Electric Motors

Use totally enclosed motors in environmentally controlled buildings, feed processing rooms, and on any equipment subject to dust and moisture accumulation. Use totally enclosed, air-over motors for ventilating fans (the air stream cools the fan).

Provide overload protection for motors in addition to the circuit breakers or fuses. Also, install a fused switch with time delay fuse at each fan and size at 125% of the motor full load current. Wire each ventilating fan on a separate circuit so if one circuit fails, fans on other circuits will come on.

If all loads on a circuit operate simultaneously, size the branch circuit for 125% of capacity based on the sum of the full-load current of each load. Long term use of a circuit at 100% capacity may cause overheating.

Size conductors to an individual motor at 125% of the motor full-load current. Size conductors to multi-speed motors at 125% of the highest full-load current rating shown on the motor nameplate. If one conducter supplies two or more motors, size it for 125% of the full-load current of the largest motor plus the sum of the full-load current of all other motors.

Service Entrances

If possible, place the service entrance in a dry, clean office or building annex. If in an animal room, use a weatherproof molded plastic enclosure and mount on asbestos board. Leave a 1" air space between service entrance and exterior wall. Install all cable entries at or near the bottom of the service entrance box, to prevent condensation from running down cables into the box. Seal cable entries to keep out moisture. Never recess the service entrance box into an outside wall, because water vapor may condense on electrical components.

Determine the required service entrance size based on present needs. Then increase it by 50% to allow for more electrical equipment later. Contact your electrical power supplier when building plans are firmed up but before construction begins.

Standby Power

Standby generators reduce the threat of power outages. A standby power system requires:

- Generator to produce alternating current.
- Engine to run the generator.
- Transfer switch.
- Alarm or automatic equipment to turn system on when power fails.

Decide on either a full- or a partial-load system. The full-load system must have capacity to carry the maximum load as well as the peak starting load. A full-load system is common if standby service is automatic. A partial-load system carries only enough load to handle vital needs and is usually controlled manually.

Tractor-driven generators are satisfactory for most operations and are generally the most economical. A problem during heavy snowfall is getting the tractor to the generator in time. If short duration outages are critical, consider an engine-driven unit with automatic switching. Operate engine-driven units for short periods at least once a week to ensure that they will function when needed. Operate tractor-driven units briefly every month.

Install a double throw transfer switch at the main service entrance just after the meter so the generator is always isolated from incoming power lines. This keeps generated power from feeding back over the supply lines, eliminates generator damage when power is restored, and protects line repairmen.

Sizing a standby power unit is difficult. Most electric motors require more power to start than to run, which further complicates the task. Assess vital power needs for a partial-load system. One procedure sums the starting wattage of the largest motor, running wattage of all other motors, name plate wattage of essential equipment, and wattage of essential lights.

Standby electric generators are a long term investment, so include power requirements of major electrical equipment purchases anticipated over the next few years. A tractor or standby engine needs a power output rating of about 2.5 times the maximum kilowatt (kw) output of the generator. Generators must operate at a constant speed, so equip the engine with a tachometer and governor.

If you have questions about standby generators, contact your local power supplier.

Water Supply

Determine the animal water needs from Table 32, then include your estimate of the other water demands. These other demands—sanitation, lagoon dilution, waterer spillage, cooling sprays, and fire protection—may be large. Water consumption depends on swine size, activity, diet, lactation, and season of the year.

Water should be of drinking quality, especially for small animals. Many states have public laboratories that analyze water samples free or for a small fee.

Table 32. Water requirements of swine.
Includes drinking and some wasted water.

32a. Individual animals.

Animal type	gal/hd/day
Sow and litter	8
Nursery pig	1
Growing pig	3
Finishing pig	4
Gestating sow	6
Boar	8

32b. Swine operations.
Based on 16 pigs sold/productive sow-yr. Manure production per productive sow accounts for all animals in the operation such as boars and her pigs in nursery, growing, etc.

Operation	gal/day
Feeder pigs produced, sold at 50 lb	10/productive sow
Pigs fed 50 to 220 lb	3/pig capacity
Farrow to finish	35/productive sow

Consider intermediate water storage, especially if standby power is not available. With low yield wells, a small pump can stockpile water during non-peak periods. Fire protection requires large quantities of water to be available quickly. Stockpile a 2 hour supply of at least 10 gallons per minute (gpm)—preferably 50 gpm.

Plumbing in environmentally controlled buildings encounters special corrosion problems. Copper pipe has a short life in swine buildings. Galvanized or black iron pipe also corrodes, especially at exposed threads. Apply a protective covering at pipe threads. PVC, CPVC, or polyethylene pipe with nylon fittings and stainless steel fasteners are best.

Plumbing can be overhead in environmentally controlled buildings if water cannot drip on electrical equipment. Overhead pipes in naturally ventilated buildings require heat tape and insulation. Insulation alone usually will not prevent freezing in open-front buildings. Drain water systems in unused buildings. Use antifreeze to protect traps that cannot be drained.

Buried lines solve most freezing problems but are very difficult to repair if they freeze. Avoid shallow water lines under outside concrete slabs. Frost penetration varies from 3′ to over 6′ in the Midwest. Frost penetrates up to 2′ deeper in compacted soil, such as under driveways and animal traffic areas, and in areas where the insulating snow cover is quickly blown away or removed. Bring underground pipes to the surface in animal pens through a larger diameter rigid plastic pipe to protect them from abuse. This also prevents damage from surrounding concrete.

Use built-in heating or continuous water flow for waterers in open-front or outside lot facilities. Place a pig-proof cover over all heaters and heat tapes to prevent equipment damage and animal electrocution.

Lightning

Install lightning arrestors to control voltage surges on electrical cables. Arrestors are particularly important on wiring to submersible pumps.

Lightning protection for buildings is usually a system of lightning rods with metal conductors to ground. Install lightning rods 20′ o.c. and within 2′ of the ends of a gable roof. Place rods at both the ridge and the eaves of buildings with a 4/12 slope or flatter roof. Fasten the metal conductors to roof and walls at 3′-4′ intervals. Drive the ground rod at least 8′ into moist earth.

Metal clad buildings that have a continuous connection between roofing and siding can be partially protected by grounding the siding unless there is flammable insulation directly beneath the roofing. However, most insurance companies require lightning rods on metal clad buildings as well. Install 2 grounding cables (on opposite corners) on metal clad buildings up to 250′ long—add another cable for each additional 100′ length or fraction. Make sure your lightning protection installer has qualified for Underwriter's Laboratories "Master Label" designation.

Incinerators

An incinerator can dispose of small dead swine—large dead swine are sent to a renderer or buried 6′ deep. Some states prohibit animal burning. Other states require afterburners on incinerators to reduce air pollution. Check with local authorities for regulations in your state.

Keep the incinerator and smokestack in good repair. Use a spark arrestor and remove dry grass and other combustible material from nearby. Locate an incinerator at least 100′ from a building and 40′ from the fuel supply.

SWINE HANDLING

The faster and quieter hogs are sorted, treated, and loaded, the less weight gains are affected. If possible, sort a few hours before loading to allow hogs to calm down. A good handling facility includes a holding pen, a sorting alley and gates, a squeeze chute or other treatment space, and a loading chute. Other features such as scales and crowd gates are also useful. Fig 79, 80, and 81 show good handling facilities.

Provide sick pens in each building for about 1% of animal capacity. Allow about 8 to 10 ft^2 / 100 lb of swine in sick pens. Locate sick pens near access doors or holding areas for convenient servicing.

Loading chutes are often the bottleneck in handling facilities. Pigs are reluctant to walk on dirt or wood floors if they were raised on concrete. If the pigs were raised on concrete floors just before shipping, consider a concrete floor for the chute. Keep chute floor clean and dry for good footing. Install steps (4″ rise per 12″ run preferred) or cleats 8″ o.c. Construction must be sturdy because pigs are reluctant to walk up a shaky chute.

Restrict chute width to 22″ or less to reduce turnarounds. Unloading chutes can be 4′-5′ wide. Slope chutes no more than 20° (about 4½′ rise per 12′ run). Pigs load better if the chute sides are solid and they can enter the truck from a flat area rather than off a

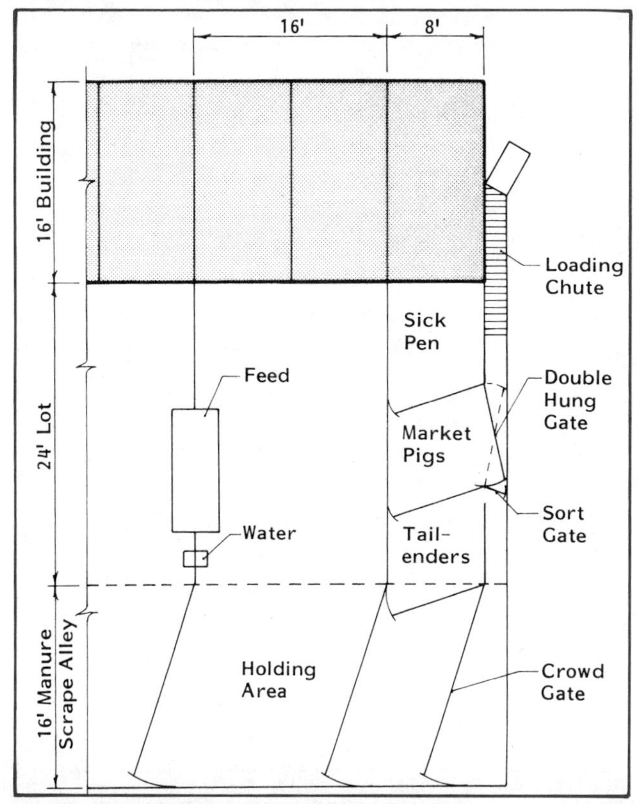

Fig 79. Handling layout for shed and lot.
An extra pen is required for sorting and sick pigs.

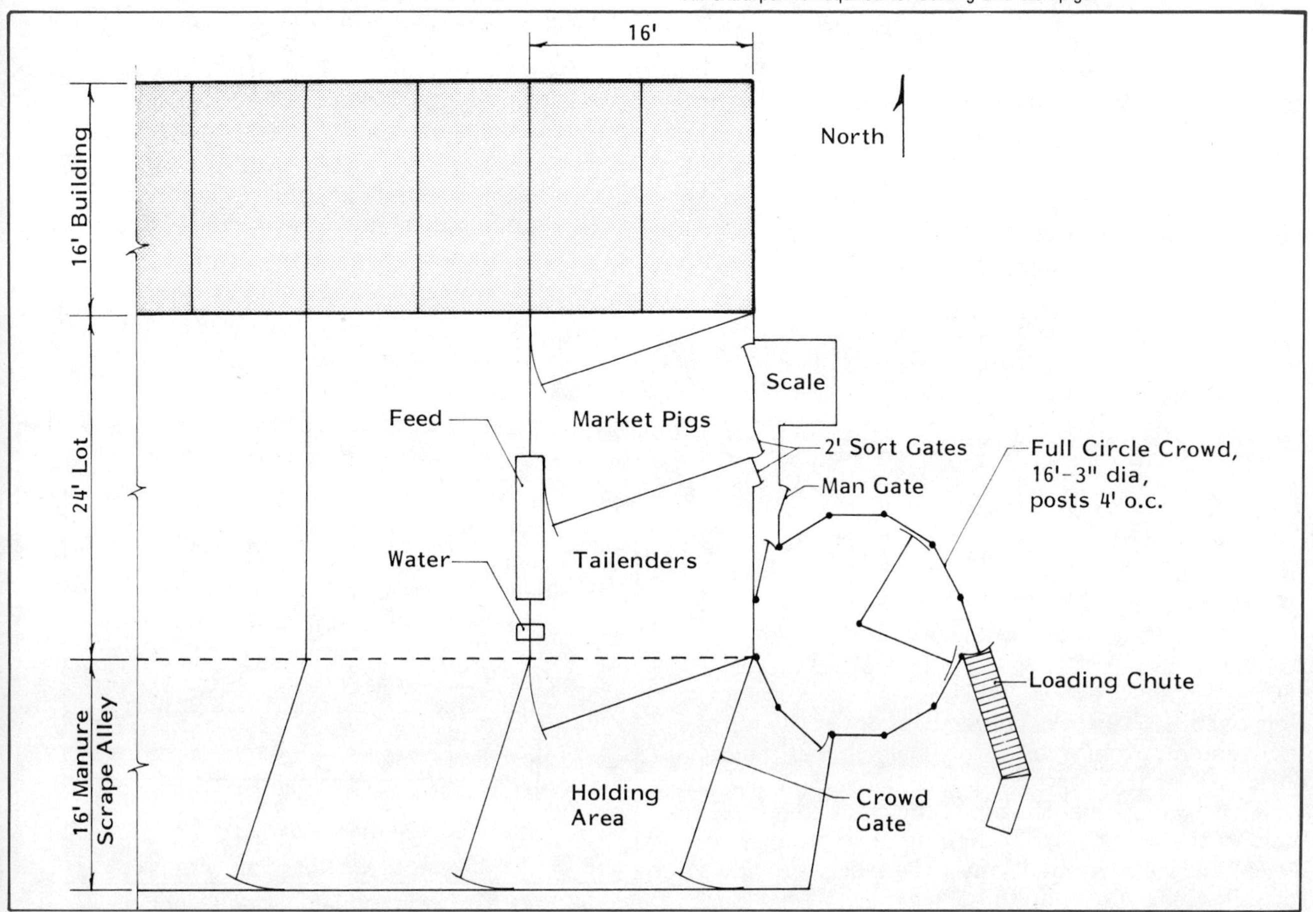

Fig 80. Handling layout for shed and lot.
End feeding pen is used for sorting. Shut up feeding pigs in building before bringing in the sorting pigs.

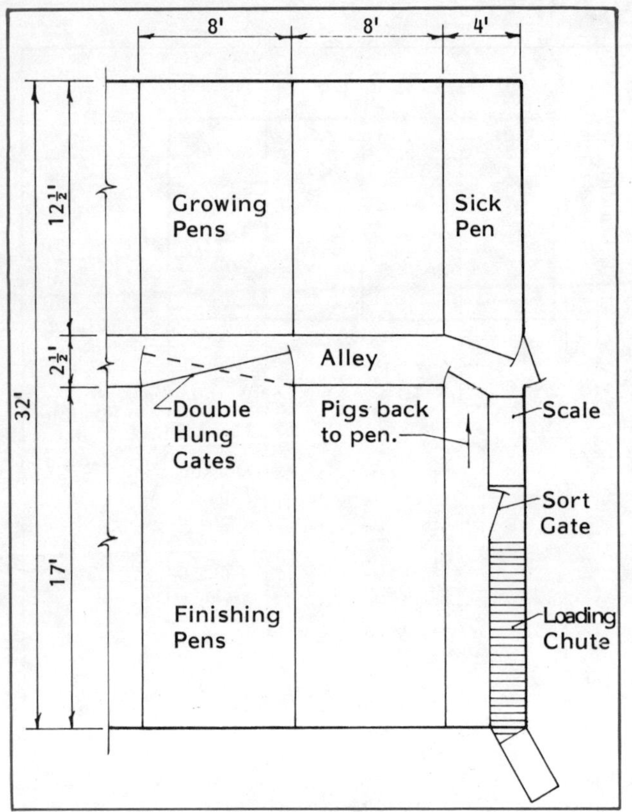

Fig 81. Handling layout in a building.

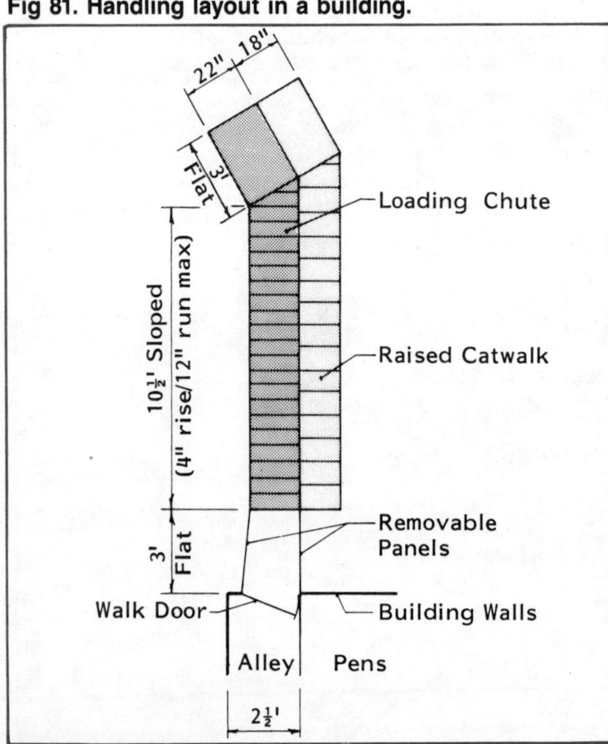

Fig 82. Permanent loading chute.

sloping chute. Turn the chute about 45° at the top of the ramp and construct a flat floor for 3′-4′ before hogs enter truck, Fig 82.

A one-way gate at the top of the chute keeps animals in the truck. A self-aligning dock bumper reduces injuries as animals enter the truck. Provide a catwalk along the chute to help you keep hogs moving.

Hogs don't like to go into bright sunlight from a dark building or vice versa. Provide uniform and diffuse light (no bare light bulbs) with no shadows across the handling area.

Build handling fences and gates (in the chute and working alley) at least 42″ high (32″ if used only for finishing hogs) with 3″ clearance between the base of the fence and the floor to prevent manure accumulation. Make the working alleys 16″ wide and use solid panels so hogs can only see and move straight ahead. Construct the circular crowd gates with open-wire hog panels, because hogs are reluctant to enter a pen with solid walls. Likewise, avoid making sow washes and other handling areas with three solid sides. Hinge sorting gates (2′ wide) at both ends to allow sorting of both incoming feeder pigs and outgoing market pigs.

Pigs can be killed or crippled in transport, usually due to overcrowding. Allow $1\frac{1}{4}$ ft^2 per 40 lb pig, $1\frac{1}{2}$ ft^2 per 60 lb pig, $3\frac{3}{4}$ ft^2 per 200 lb pig, and $4\frac{1}{4}$ ft^2 per 250 lb pig on trucks.

Livestock Conservation, Inc., suggests the following guidelines for transporting swine.

- Make trailers and loading facilities free of protruding hardware and boards.
- Load as quickly and as quietly as possible and leave immediately.
- Check animals about one hour after loading, then keep moving to reduce travel time.
- Provide good footing to reduce injuries.
- Provide bedding—use straw on dry sand if below 60 F; wet sand if above 60 F.
- Use partitions to avoid piling due to sudden stops or cold weather.
- Keep vehicles as open as possible in summer to maximize air movement over animals. Keep vehicles as closed as possible in winter.
- In summer, load and haul during the cool of the day.
- If the temperature exceeds 80 F, wet down animals and increase floor space per pig by one-third.
- Put 25 lb ice blocks on truck floor in very hot weather.

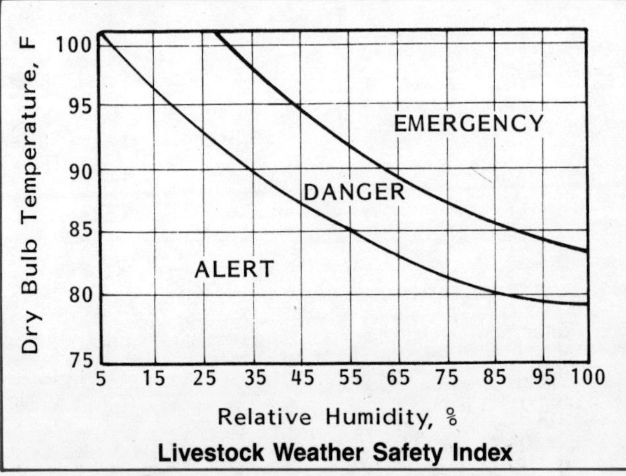

Fig 83. Weather guide for safe swine shipping.
From Livestock Conservation, Inc.

EXISTING BUILDINGS—REMODEL OR ABANDON?

When planning to expand, you may have to decide whether to remodel or abandon an existing building. Carefully consider your future, as well as current, needs. Conditions vary from farm to farm, but always evaluate these general factors.

- Compatibility of existing building to final setup.
- Structural integrity of existing building.
- Location of existing building.
- Cost of remodeling vs. new building.

Compatibility

The "best" remodeled building is one that requires few modifications. If an existing building must be changed drastically to meet your current needs, consider a new building.

Alternative Uses

First, decide if the existing building is more like another building you are planning to build in the future, rather than the building you are planning now. For example, it is easier to remodel an uninsulated machine shed for breeding-gestation than for a nursery or farrowing facility. Therefore, consider a new building for a nursery now. Remodel the machine shed into a gestation building later.

Factors affecting this decision include ceiling/roof height, location, adaptability to mechanical or natural ventilation, and insulating ease. Farrowing and nursery buildings require ceiling heights of 8′ or less, mechanical ventilation, and good insulation. Finishing hogs and gestating sows produce well in buildings with 8′-10′ high ceilings, natural ventilation, and less insulation.

Two-story barns work better for farrowing and nurseries, because they are usually well insulated with low ceilings. They do not adapt well to natural ventilation. Machine sheds work better for finishing and gestation, because they are easier to modify for natural ventilation and light insulation.

Interior Space Arrangement

Determine if the existing layout adapts easily to the desired floor plan. Post locations frequently do not fit the new plan. It is possible to span existing posts with sub-beams or new posts and beams, but relocating a post usually requires a new footing. If you plan to use an existing concrete floor, new footings are costly and time-consuming.

Remodeled buildings usually make poor use of space—too much space in some areas and not enough in others. For example, a 20′x58′ building remodeled for farrowing in 20-5′x7′ stalls has only a 2′ center alley but 2′-3′ left over at one end. The narrow center alley is crowded and inefficient, and when considered over the life of the building, may make a new building desirable.

Environmental Control System

Determine if a good ventilating and heating system can be installed. For example, it is impractical to install floor heat in an existing concrete floor, so the building is not a good candidate for a farrowing building. Consider not only the initial cost of ventilating equipment, but also the cost of operating it in a less efficient remodeled building compared to a new building.

Check if air inlets and fans can be easily installed and properly located. See section on ventilation. Problems include:

- Ceiling obstructions. Exposed joists and beams block proper air distribution. A ceiling or fan and tube inlet system may be required.
- High ceilings. You may need a lower "false" ceiling to reduce drafts and heated space.
- Unplanned air inlets. Exhaust ventilation does not work well in a leaky building. Positive or neutral pressure ventilation may be required.

Manure Management

Will your manure handling system work in the remodeled building? It is difficult to add manure storages to existing buildings. If a flushing or scraping system is unsuitable, remodeling may be inadvisable. Special equipment to handle manure just for the remodeled facility is part of the remodeling cost.

Rodent Proofing

Can the existing building be made reasonably rodent proof? A flat slab or shallow footing (less than 2′ below grade) is inappropriate for livestock facilities, because rodents burrow and nest under the foundation.

Structural Integrity

Carefully evaluate the building's structure, whether it is for short- or long-term use. Check alignment and condition of foundations, strength and alignment of sidewalls, and condition of roof framing and covering. A building with a swayback ridgeline may have been overloaded or poorly built, and could need high future maintenance or repair costs. Minor roof leaks may be acceptable in growing-finishing or breeding-gestation buildings, but they are unacceptable in insulated farrowing or nursery buildings. Repairing a foundation is expensive and difficult.

Location

Evaluate location not only for suitability to the remodeled building but also for future buildings planned for the same site. Consider:

- Farm home and neighbors. If the existing building is upwind from you or your neighbor, question converting it to a swine facility.

- Runoff and drainage. It may be impossible to maintain warm, dry floors in a building in a low area.
- Manure handling. Space must be available for manure handling facilities such as lagoons.
- Access roads. Does the present road provide all-weather access for heavy vehicles? Is building access convenient; can it be supervised?
- Room for expansion. Consider future buildings, feed centers, access roads, and manure handling facilities.

See section on site selection.

Do not compromise on location when deciding whether to remodel a building. A common argument is that the location may not be good, but the remodeled building is only temporary. After spending several thousand dollars to remodel a building, it often becomes the focal point for all future livestock building and feed handling construction—all in the wrong location.

Economic Considerations

Remodeling is not always the cheaper route, especially when future needs are considered. If remodeling cost is more than ½ to ⅔ of the new building cost, a new building is usually best. It may be possible to use some materials from the existing building for the new building. Remodeling costs include demolition of interior structural components and concrete floors.

Consider cost and availability of construction labor. Due to the many "unknowns" in remodeling, many farm builders are hesitant, if not unwilling, to accept remodeling jobs. If they accept a remodeling job, the contract is generally so heavily "padded" to compensate for possible "unknowns" that labor becomes very expensive.

Do not plan a new $100,000 swine facility around an existing $5000 building or $1000 concrete slab. Existing facilities may save money initially, but the long-term cost of restricted expansion and vehicle movement and reduced labor efficiency may offset the initial savings.

GRAIN-FEED CENTERS

Layout

Ideal grain flow design develops a closed loop for handling each storage and processing procedure. Harvested grain is received in the grain dump, conveyed to drying, and moved to storage. The loop is completed when the bin is unloaded through an underfloor conveyor to the grain dump for loadout, for transfer to another bin, or for feed processing. Sometimes grain bins unload directly into a portable grinder-mixer or truck.

If you have a drive-over dump and vertical elevator, make them the focal point of your layout. Add storage bins around this core, Fig 84.

Gravity conveying is convenient, but highest in cost, so use only where its special characteristics are essential to the process. Overhead bins require extra supports and taller elevators.

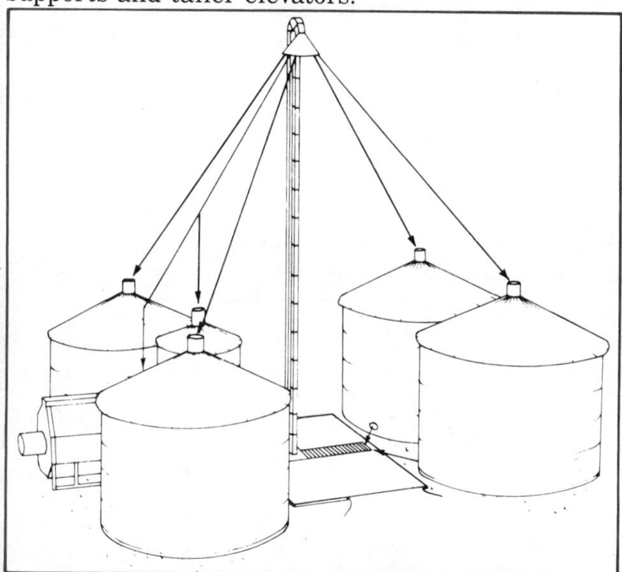

Fig 84. Several loops built around a drive-over dump.

Grain Storage

The biggest bin is not always the cheapest per bushel storage. Generally, the biggest bin should not exceed half of the total volume of the operation. For freshly harvested grain, equip storage bins with aeration to cool the grain. For more information on grain aeration, obtain *Managing Dry Grain In Storage*, AED-20, by Midwest Plan Service.

Grain for swine can be stored either wet or dry. Both storage methods have about the same overall costs. Differences are in the ratio of investment cost to operating expense. Drying and storing shelled corn takes a moderate capital investment, but may have high operating expenses. Capital investment for high moisture grain can be very high, but operating expenses are much less than for dry grain.

High Moisture Feed Ingredients

High moisture cereal grains are suitable swine feed either ensiled or preserved with organic acid treatment.

Ensiling

Sealed storage is the most convenient for high moisture grain. It increases handling ease and eliminates most of the spoilage found in unsealed systems. Special oxygen-limiting silos are required for sealed storage. Grain need not be cracked or ground before storing, but is usually passed through a roller mill before feeding. Sealed storages often unload from the bottom.

Unsealed high moisture grain storage is less common for swine feed. Grind the grain before storage to ensure adequate packing to exclude oxygen. Ideal particle size is similar to medium ground dry grain. Careful management and feeding at least 3" per day from the top are required to minimize spoilage.

Preservation With Organic Acid

High moisture grain, especially corn, can be treated with organic acid for storage up to one year without damage or loss of feeding quality for swine. When used correctly, acid preservatives are not toxic to swine and do not affect feed palatability. The treatment cost per bushel may equal or exceed drying costs. Use nonmetallic contact surfaces in storages to avoid corrosion by the acid.

Feeding Methods

Feeding methods are essentially the same for all high moisture cereal grains. Grain is fed alone, top dressed with supplement, or part of a complete mixed feed. Grinding, rolling, or cracking wet grain does not improve animal efficiency. However, grinding does improve mixing of complete rations. There are two feeding methods for high moisture corn:

- Complete mixed ration ad libitum feeding.
- Limit feeding on floors.

Ensiled high moisture grain spoils shortly after removal from storage. Wet grain left in conveyors for 2 to 3 days also tends to spoil.

To keep high moisture grain fresh in the feeder:

- Feed 2 to 3 times a day at temperatures above 80F.
- Feed daily at temperatures below 80 F.

Because little feed can be stored in feeders, be prepared for power failure or equipment breakdown.

Dry Feed Ingredients

Storage

Dry feed ingredients are stored in:

- Ground level, flat-bottom bins with unloading augers for dry grain. Size for a two-week to one-month feed supply, or for a one-year supply of on-farm produced grain.
- Ground level, hopper-bottom bins, for soybean meal, complete rations, and grain required in small amounts. These bins cost more per ton capacity than flat-bottom bins. Materials that flow poorly, such as freshly ground soybeans, require a "live-bottom" bin to reduce bridging.

Completely empty hopper-bottom bins periodically.

- Overhead hopper-bottom bins provide gravity flow to processing or mixing equipment. They are expensive—use only where they reduce labor. Do not use overhead bins for storage.

Provide one to two weeks storage for farm processed feed rations. Make allowances for bad weather, delayed deliveries, and rush seasons, but keep feed fresh. Compare price advantages for timely, seasonal, or volume purchases against the storage cost. Volume purchases of grain and soybean meal may reduce the number of adjustments required in rations. Adjustments have to be made when moisture, protein, energy, fiber, etc., change appreciably. However, degradation of certain vitamins require that pre-mix and base-mix materials be fed within 90 days of manufacture. Match volume purchases and storage capacity to usage.

Feeding Methods

Dry feed is usually self-fed as a complete mixed ration. Size feeders for no more than 4 days storage for maximum palatibility. Increase feeding frequency in warm weather and in farrowing and nursery rooms at 70 F or higher. Floor feeding requires more management for maximum productivity without feed waste.

Feed Processing

Unloading

Unloading grain from storage for feed processing is a low capacity operation. If grain storage is also used for cash grain, high volume unloading equipment is needed. Provide a surge bin at the mill to eliminate frequent starting and stopping of unloading equipment.

Processing

Feed processing involves grinding grain, weighing or metering ingredients, mixing ingredients, and delivery to bulk tanks or feeders. Fig 85 illustrates basic systems, and Table 33 summarizes system features.

With batch feed processing, individual ingredients are weighed, ground or rolled (in the case of wet grain), and mixed in batches. Farm batch systems are usually not totally automatic, although some steps can be automated, such as grinding grain into a holding bin and automatic delivery of feed batches.

With continuous-flow feed processing (very high initial cost), grain is milled and blended with other rations automatically. Switches stop the mill when the supply of any ingredient is exhausted or when the bin to which feed is being delivered is full.

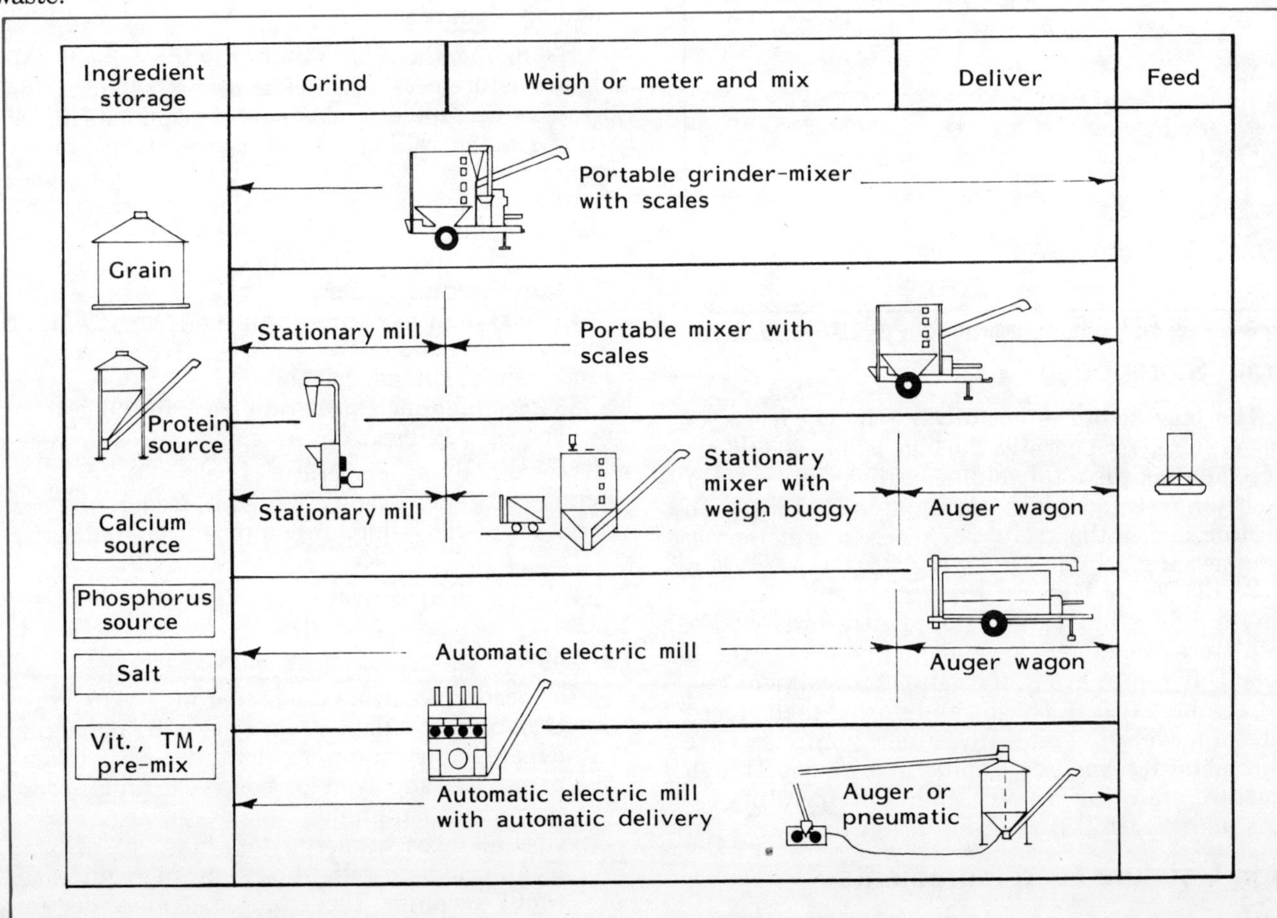

Fig 85. Alternative grinding-mixing systems.

Table 33. Summary of grinder-mixer features.
Actual break-even volume depends on type and capacity of supporting equipment storage etc. From Pork Industry Handbook-4, "On Farm Feed Processing."

Type of system	Usual break-even volume	Advantages	Disadvantages	Comments
Portable grinder-mixer	200-400 tons/yr (30-60 sows)	• Portable—can pick up and deliver at existing facilities • Can grind hay	• Highest labor • Highest operating cost • Requires a tractor	• The most popular first system • Purchase of a used unit allows inexpensive start of on-farm processing
Mill with portable mixer	200-400 tons/yr (30-60 sows)	• Can pre-grind grain • Pick up and delivery at existing facilities	• High labor • High operating cost • Requires tractor or truck	• Mill may be stationary or portable • Mill can be located at grain storage
Stationary mill with stationary mixer	200-400 tons/yr (30-60 sows)	• Excellent control of ration composition • Can pre-grind grain • Can include automatic delivery	• High labor • Components must be matched • May not be able to use existing facilities	• Favored by producers who want direct control over ration composition
Automatic electric mill	200-400 tons/yr (30-60 sows)	• Lowest labor • Lowest operating cost • Long service life • Can include automatic delivery	• Must be routinely calibrated • Will not handle hay • May not be able to use existing facilities	• Planning is required to make best use of automatic features
Package feed center with automatic mill	800-1,200 tons/yr (100-180 sows)	• Pre-engineered for reliability • Quick installation • Bucket elevator can serve surrounding storage • Can include automatic delivery	• High cost due to overhead bins • Few options in size of unit	• Additional space for sacked ingredients and ground-level storage of grain and soybean meal is normally required
Batching plant-highly mechanized	2,000-4,000 tons/yr (300-600 sows)	• Best ration control for high volume systems • Calibration not required	• First cost requires high volume • Must be custom built	• Careful design is required for most efficient operation

Portable Grinder-Mixer

Portable grinder-mixers are versatile. They can collect ingredients from a number of locations and process feed in batches. Grinding, mixing, and delivery are done by one machine. Portable grinder-mixers can be equipped to grind and handle hay.

Larger models hold about 120 bu (3 ton). Typically, it collects ingredients, grinds, mixes, and delivers a load of finished feed in about an hour.

Higher labor and operating costs are the major disadvantages of portable grinder-mixers. Both a tractor and an operator are required, and all ingredients must be weighed into the unit separately.

Select a portable grinder-mixer with features that save time and labor, such as large hoppers and high grinding capacity. Augers driven by hydraulic motors improve reliability and increase flexibility. Electronic scales are convenient, improve accuracy, and are an essential part of the grinder-mixer.

Mill With Portable Mixer

A stationary or portable mill grinds the grain in individual batches as needed. High grinding capacity is important to save time. Or, automate a low capacity mill to pre-grind grain into a bulk tank, which can rapidly load the portable mixer.

A portable mixer collects the ingredients, mixes the ration, and delivers it to bulk tanks or feeders in batches. Distances involved sometimes make a truck-mounted mixer more practical than tractor-drawn models. Equip portable mixers with electronic scales.

Stationary Mill with Stationary Mixer

A stationary mill with mixer offers good control over feed ration composition. A wagon or truck delivers the feed to bulk tanks or feeders. For large swine operations, feed is delivered by high capacity pneumatic systems (conveying in an airstream) or by high capacity overhead conveyors. Pneumatic conveyors are convenient but have high energy requirements. Mount the stationary mixer on scales to weigh ingredients, or use a weigh buggy and dump into the mixer.

Automatic Electric Mills

Automatic electric mills are reliable and accurate, if kept calibrated. They have low labor and operating costs, consuming about 3 kw-hr of electricity per ton of feed.

These mills meter the ingredients into the grinding chamber where they are mixed. These mills do not handle hay. Capacities vary from one ton/hr for 3-HP models, to 8 ton/hr for 20-HP models (roughly 600-800 lb/HP-hr).

Store grain and other ingredients close to the mill. A separate means of feed delivery is required. Automatic electric mills require planning to get full advantage of automatic operation. While mill capacity is low, many automatic mills operate 30 or more hours per week producing complete rations for large operations with little operator time.

Package Feed Centers

You can buy a complete feed center with all components pre-engineered for easy construction and reliable performance. All components are matched in capacity. Package feed systems are usually more expensive than custom designed systems. When quick construction is important or large annual feed volume uses the convenience features, package feed centers are justified.

Package feed centers are either batch or continuous-flow automatic types. Batch types consist of overhead bins, stationary grinding mill, stationary mixer, and conveyors. Continuous-flow types consist of overhead bins feeding an automatic electric mill by gravity and a vertical, finished-feed auger. The area under the bins is enclosed to form a small building to house the mill. A bucket elevator and drive-over dump can be included in the package. Bucket elevators serve surrounding grain bins as well as overhead bins.

Delivering Rations

Portable grinder-mixers, portable mixers, and auger wagons or trucks can deliver feed to bulk tanks or directly to self-feeders.

Often feed is delivered to bulk tanks at the buildings, then conveyed to feeders inside the buildings with augers, cable or chain units, or small cleated belt conveyors. Save additional labor with automatic delivery to the bulk tanks in overhead or underground conveyors.

For typical feed processing capacities, use a 6″ diameter overhead auger or a 3″ pneumatic system for batch systems. Small automatic mills can be served by 4″ augers or 1½″-2″ pneumatic systems.

Locate all buildings close enough to the mill to make automatic delivery practical. Overhead conveyors are reliable up to 300′-400′. Pneumatic conveyors deliver full volume for distances up to 800′. Beyond this distance, reduce conveying capacity or stage the pneumatic system.

Locating a Grain-Feed Center

Locate grain-feed centers near the major feed user, such as a finishing building, if conveying feed automatically to bulk bins. Place it for convenient vehicle access if hauling feed.

Consider the following:

- **Leave room for growth.** More storage, faster drying, a feed processing center, or larger vehicles may be needed in the future. Plan for maximum mechanization, even if it is added later.
- **Relationship to other buildings.** Place the feed-grain center behind or to one side and at least 200′ from the home, to reduce noise and improve appearance. Draw a map of the existing farmstead and determine how feed will be transferred.
- **Access.** Provide all-weather access roads. Arrange the facilities so transport vehicles can drive through without opening gates and without backing up.
- **Drainage.** Choose a well-drained site. Provide drainage for dump pits. Perimeter drains around footings and polyethylene vapor barriers beneath floors are essential.
- **Soil.** Have soil tested for strength to support heavy storages and equipment. Otherwise, the structures may settle.
- **Overhead space.** Avoid overhead power lines interfering with tall bins, elevator legs, or other vertical conveyors. Leave room to assemble a leg on the ground and lift it into position with a crane. Also, check for adequate clearance for leg guy wires.
- **Orientation.** Direct dryer fans away from the home for the least noise. With batch-in-bin dryers, put the bin between fan and home. Also, put the dryer where prevailing winds are least likely to carry dust or foreign material toward the house or across driveways.
- **Utilities.** Check the availability of adequate single phase or 3-phase power and/or petroleum fuel before buying any equipment. Many utilities limit single phase motors to 10 HP.
- **Construction checklist.** Have a complete plan for the grain-feed center before construction begins. Show alternatives which allow expansion or improvement of the system. Stake out the location of all facilities before construction starts.

Safety

Augers are one of the most dangerous items on the farm. Cover all auger intakes with a grate designed to keep hands, feet, and clothing from contacting the auger. Make the grate strong enough to support a man. Follow OSHA standards in equipment selection and installation.

Anyone in a grain bin when the unloading auger is running risks suffocation or injury. A knotted safety rope hanging near the center of the bin offers little protection against the tremendous pull of unloading grain. Never start machinery before locating children and coworkers.

Disconnect power to the unloading auger before entering the bin. Place an on-off switch at the bin entrance.

Spoiled or wet grain sometimes bridges over the auger and inhibits unloading. Never walk on the bridge because it may collapse and trap you. Break up the bridge with a long pole from the outside.

Wear an effective dust mask when exposed to grain dust. In particular, avoid breathing mold dust from spoiled grain. Handling milo (grain sorgham) is risky, so take precautions against inhaling its dust.

EQUIPMENT

FENCES

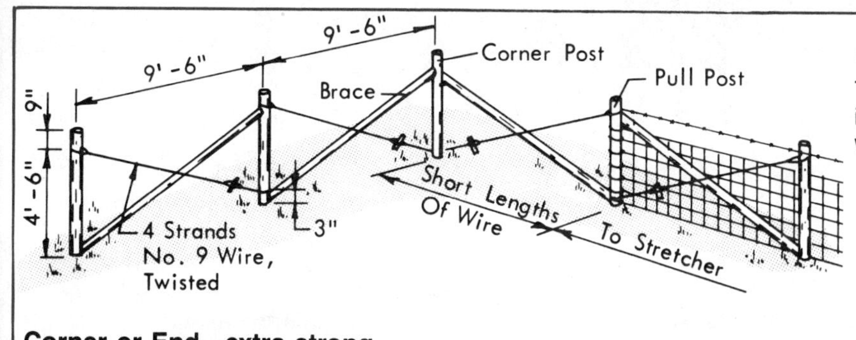

Corner or End—extra strong

Construction Steps

This extra strong fence corner, or end, is good in soft soils, or where deadman on corner post would otherwise be necessary.

1. Set all fence posts.
2. Install bracing.
3. Fasten wire to second post.
4. Tighten from second post, and complete line fence.
5. Using short lengths of wire, close corner.

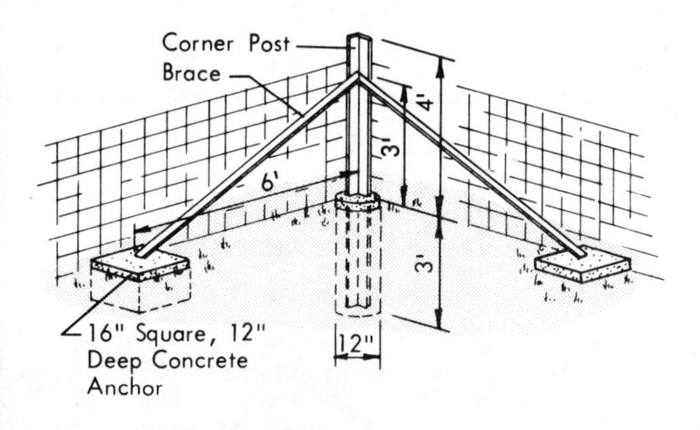

Corner or End—steel posts

Post Sizes

End Post, min sizes
- 2½" x 2½" x ¼" Angle
- 2" I.D. Standard Pipe
- 5" Top Wood Post—8' long

Brace Post, min size (wood)
- 1st Brace Post—5" top, 8' long
- 2nd Brace Post—4" top, 8' long

Brace

For Angle or Pipe Corner Post
- 1¼" I.D. Standard Pipe
- 2" x 2" x ¼" or 3/16" Angle

For Wood Corner Posts
- 2" I.D. Standard Pipe
- 2" x 2" x ¼" or 3/16" Angle

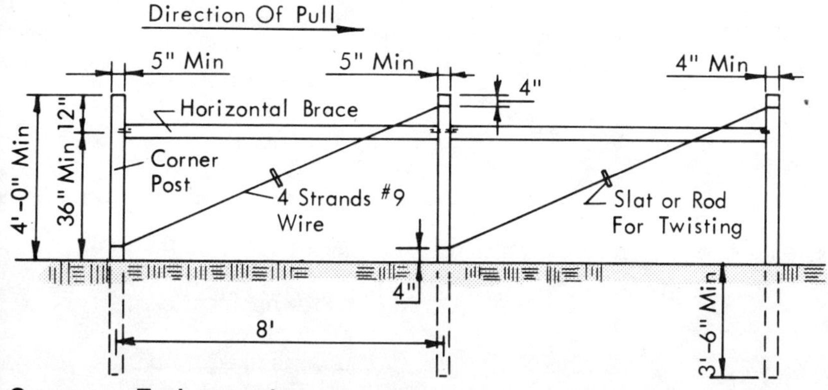

Corner or End—wood posts

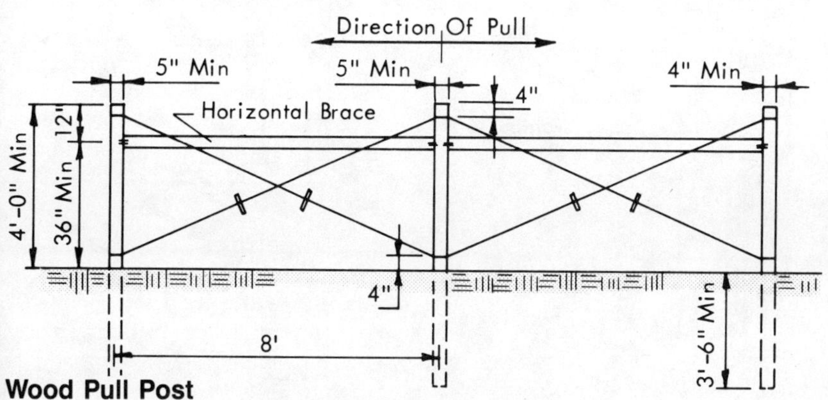

Wood Pull Post

For middle of long fence, place about 660' apart.

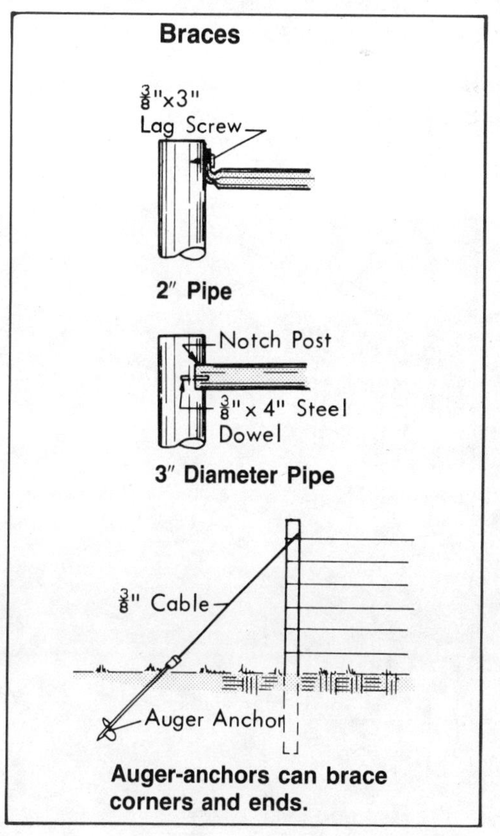

Auger-anchors can brace corners and ends.

Fences

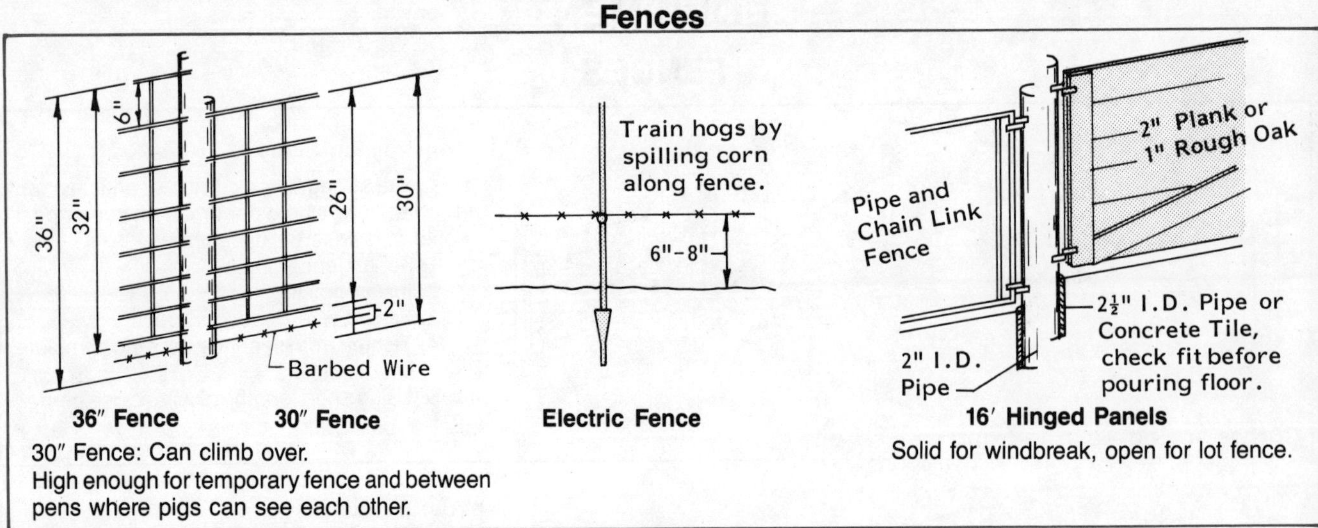

36″ Fence **30″ Fence**

30″ Fence: Can climb over.
High enough for temporary fence and between pens where pigs can see each other.

Electric Fence

16′ Hinged Panels

Solid for windbreak, open for lot fence.

Gates

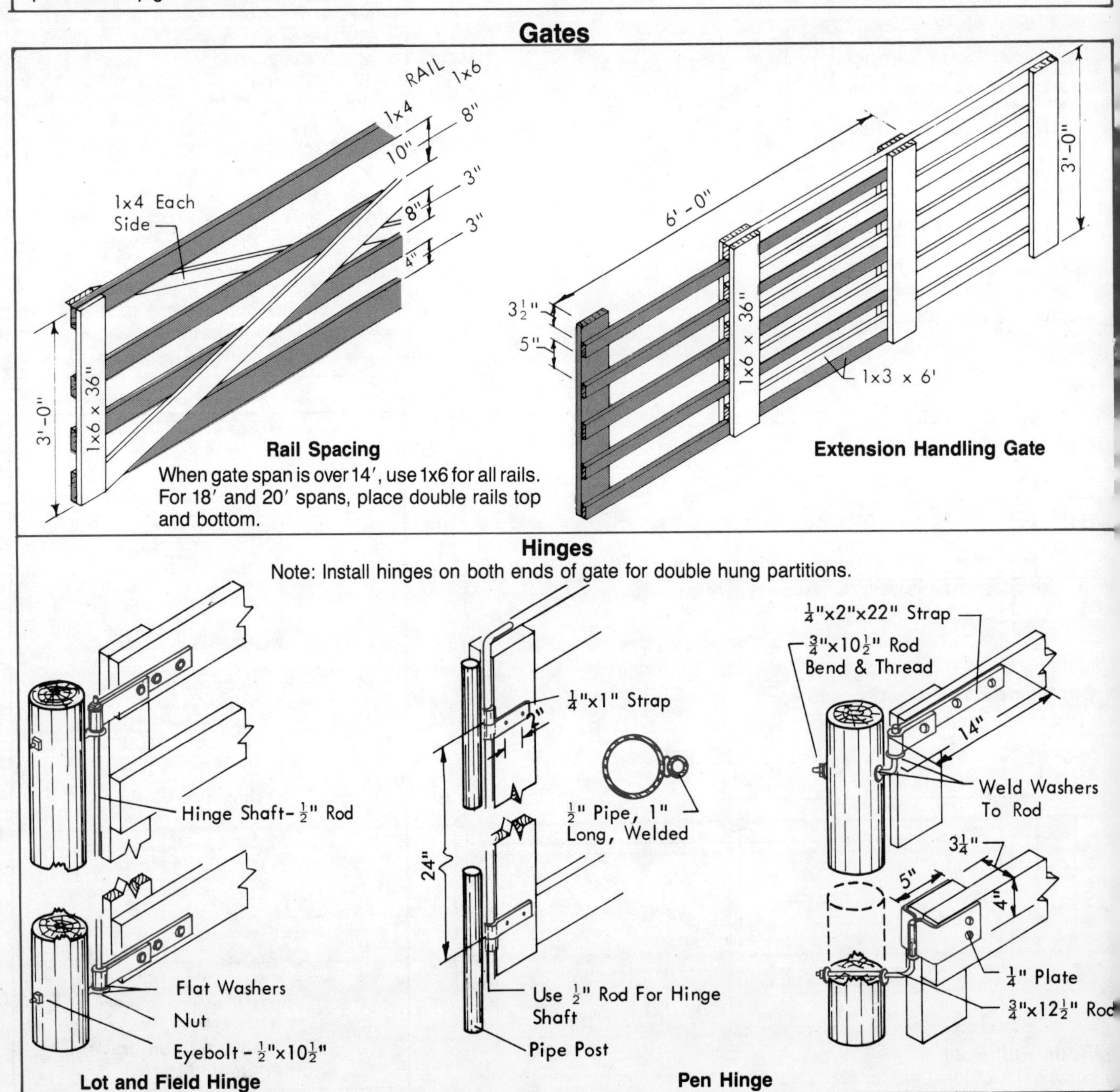

Rail Spacing

When gate span is over 14′, use 1x6 for all rails. For 18′ and 20′ spans, place double rails top and bottom.

Extension Handling Gate

Hinges

Note: Install hinges on both ends of gate for double hung partitions.

Lot and Field Hinge

Pen Hinge

Gates

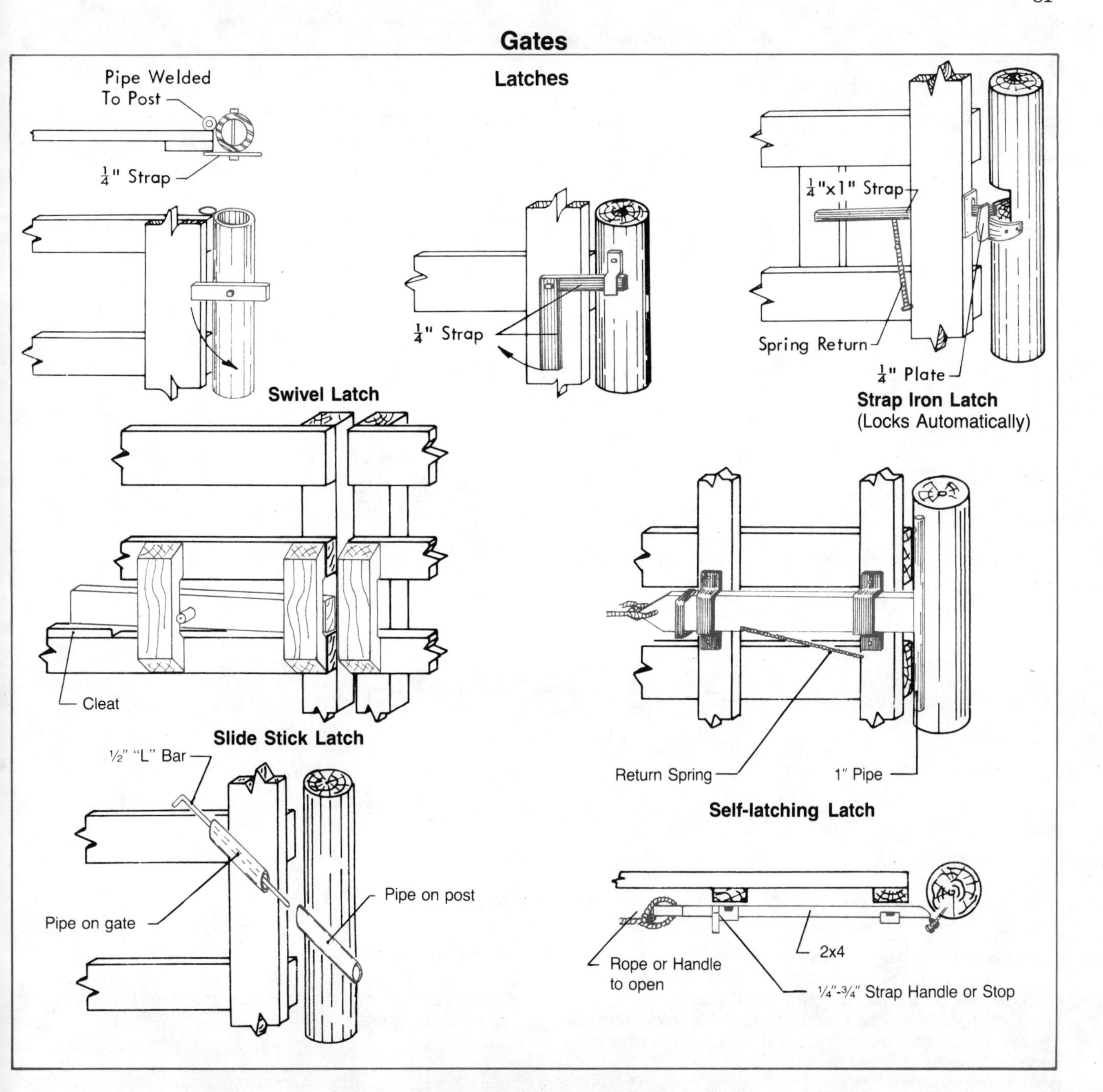

Plank Fences Splices

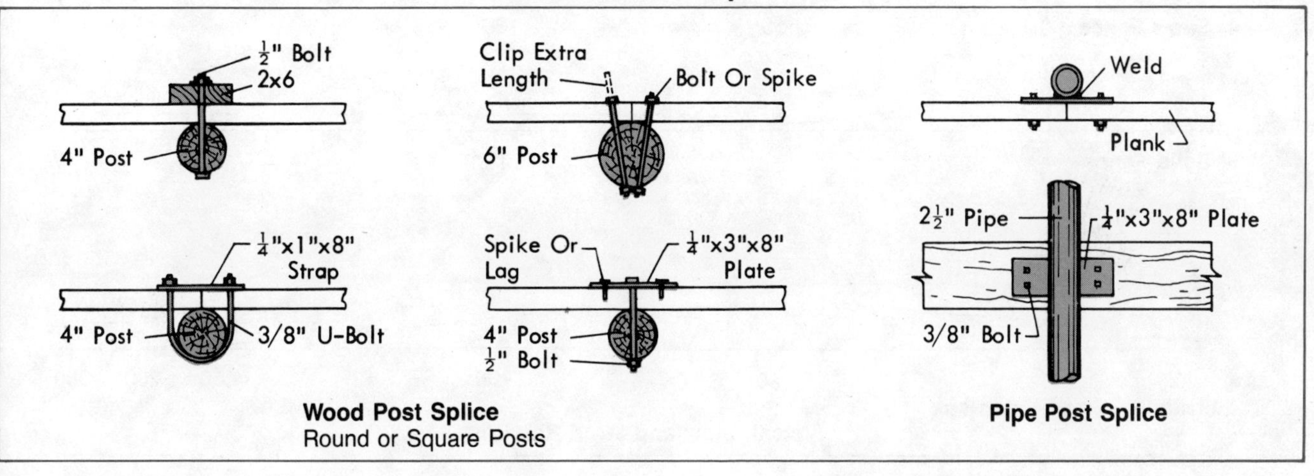

Windbreak Planning

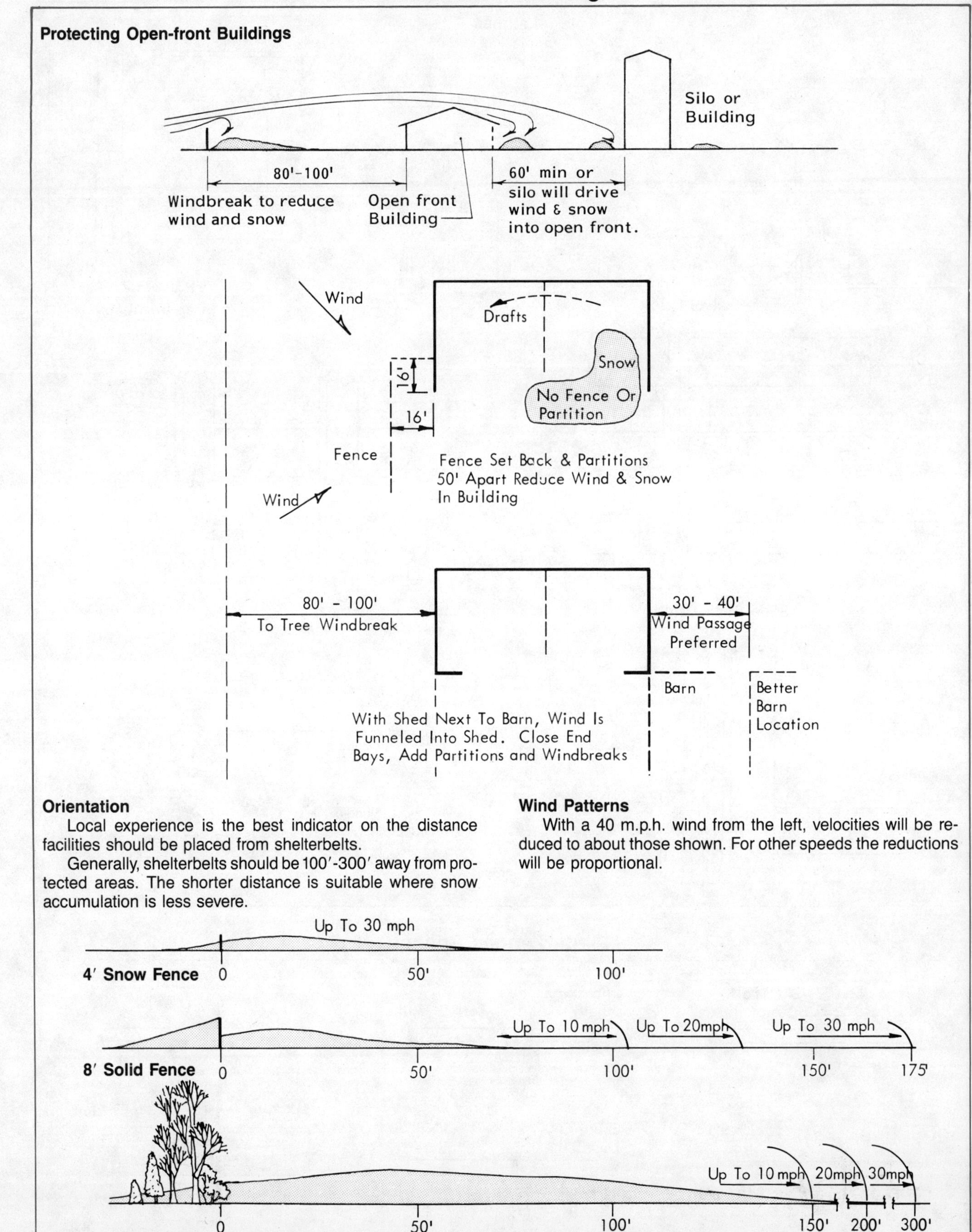

Orientation

Local experience is the best indicator on the distance facilities should be placed from shelterbelts.

Generally, shelterbelts should be 100′-300′ away from protected areas. The shorter distance is suitable where snow accumulation is less severe.

Wind Patterns

With a 40 m.p.h. wind from the left, velocities will be reduced to about those shown. For other speeds the reductions will be proportional.

Typical Snow and Wind Patterns

Windbreak Fences

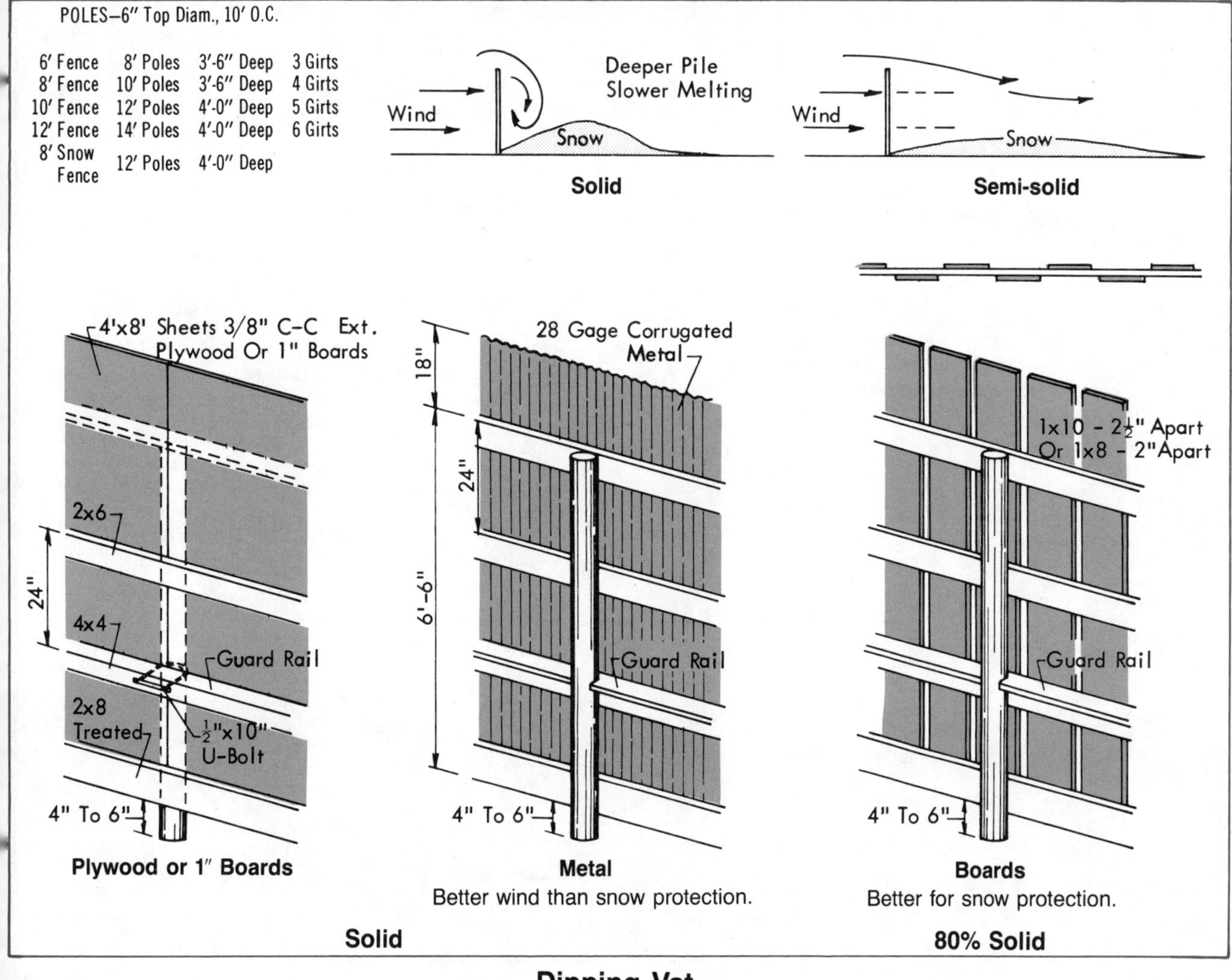

Dipping Vat

4" top x 6'-0" Post
1'-4"
24"
Approach Chute, 32" high min, solid fence
Steel Trowel Finish
Drain into Tank (6" fall)
3'-6"
6'-0"
6'-0"
19'-6"
8'-0"
12'-0"

Plain View

4" top x 6'-0" Post
24"
Grade
16"
6" Wall
6" Wall
4" Floor

Cross Section

32" min
Grade
12"
6"
24"
4'-6"
24"
30"
Liquid Level
4" Floor
4 mil Plastic over 2" Sand
6" Wall
¼" Rods, 12" o.c., both ways
Grooved surface so hogs can get footing.
Coarse Grain Backfill
Drain
8'-0" Vat
10'-0" Ramp
20'-0" Dripping Pen

Longitudinal Section

HANDLING EQUIPMENT

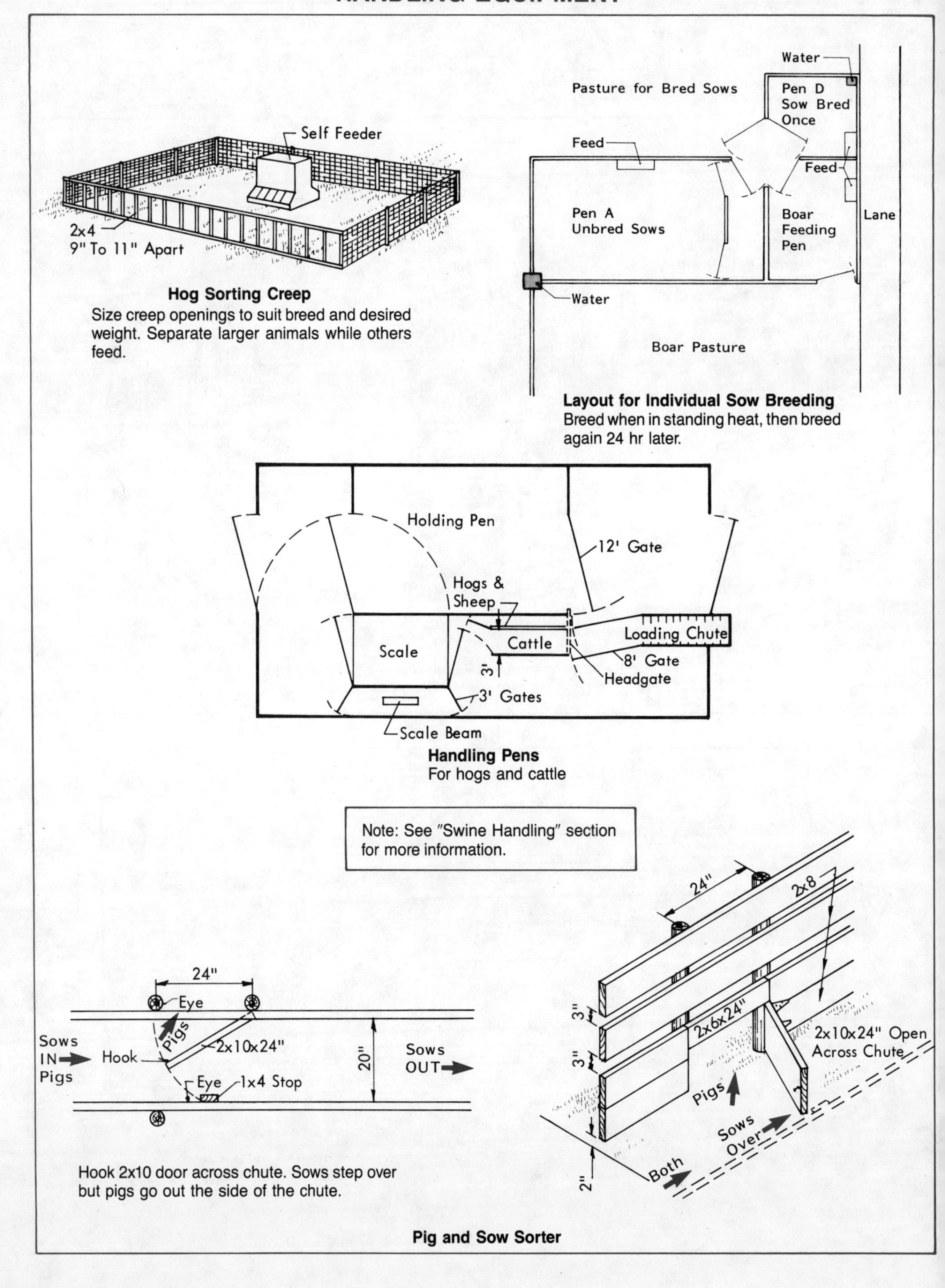

Hog Sorting Creep
Size creep openings to suit breed and desired weight. Separate larger animals while others feed.

Layout for Individual Sow Breeding
Breed when in standing heat, then breed again 24 hr later.

Handling Pens
For hogs and cattle

Note: See "Swine Handling" section for more information.

Hook 2x10 door across chute. Sows step over but pigs go out the side of the chute.

Pig and Sow Sorter

Sorting Chute

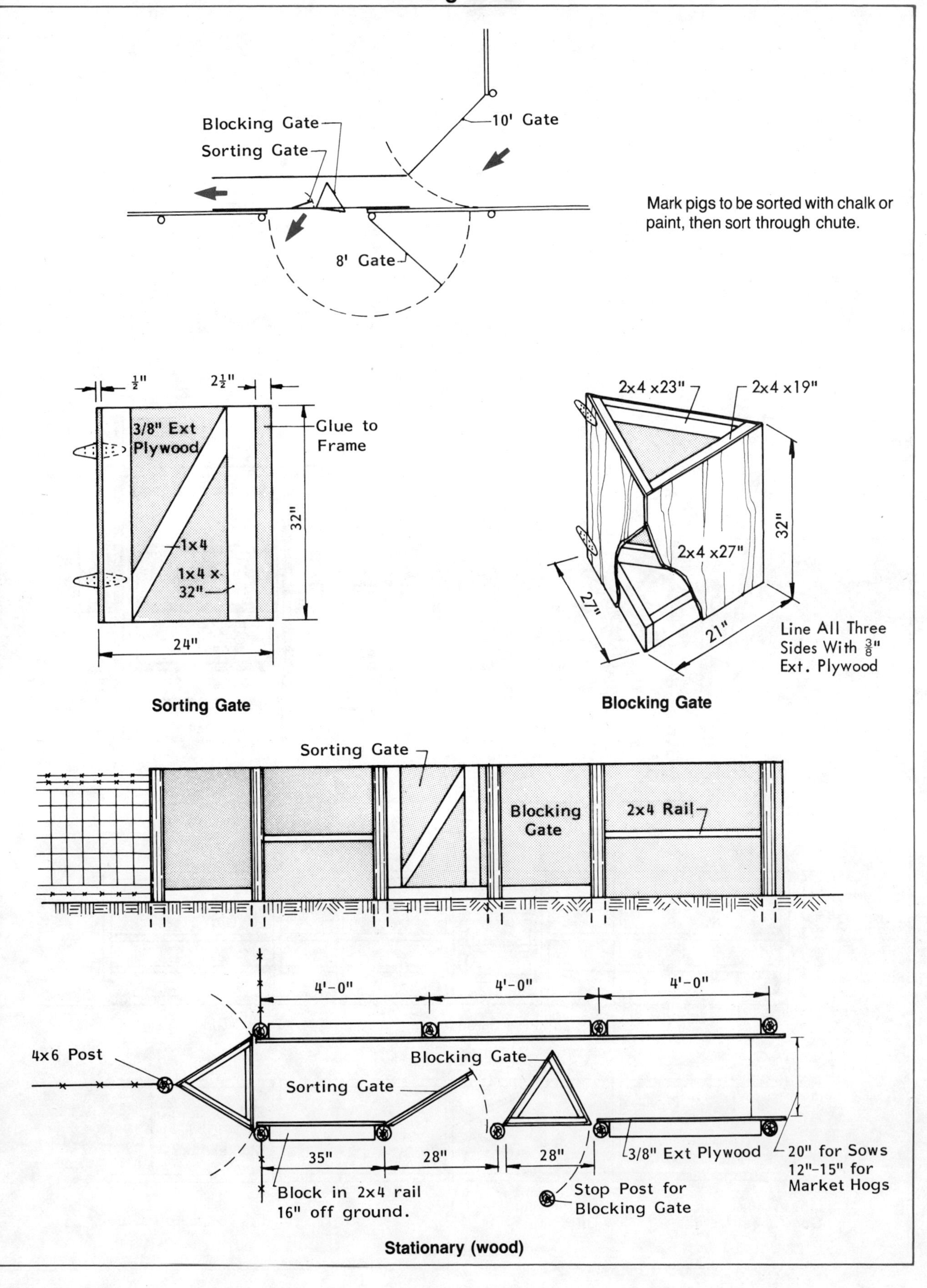

Sorting Chute

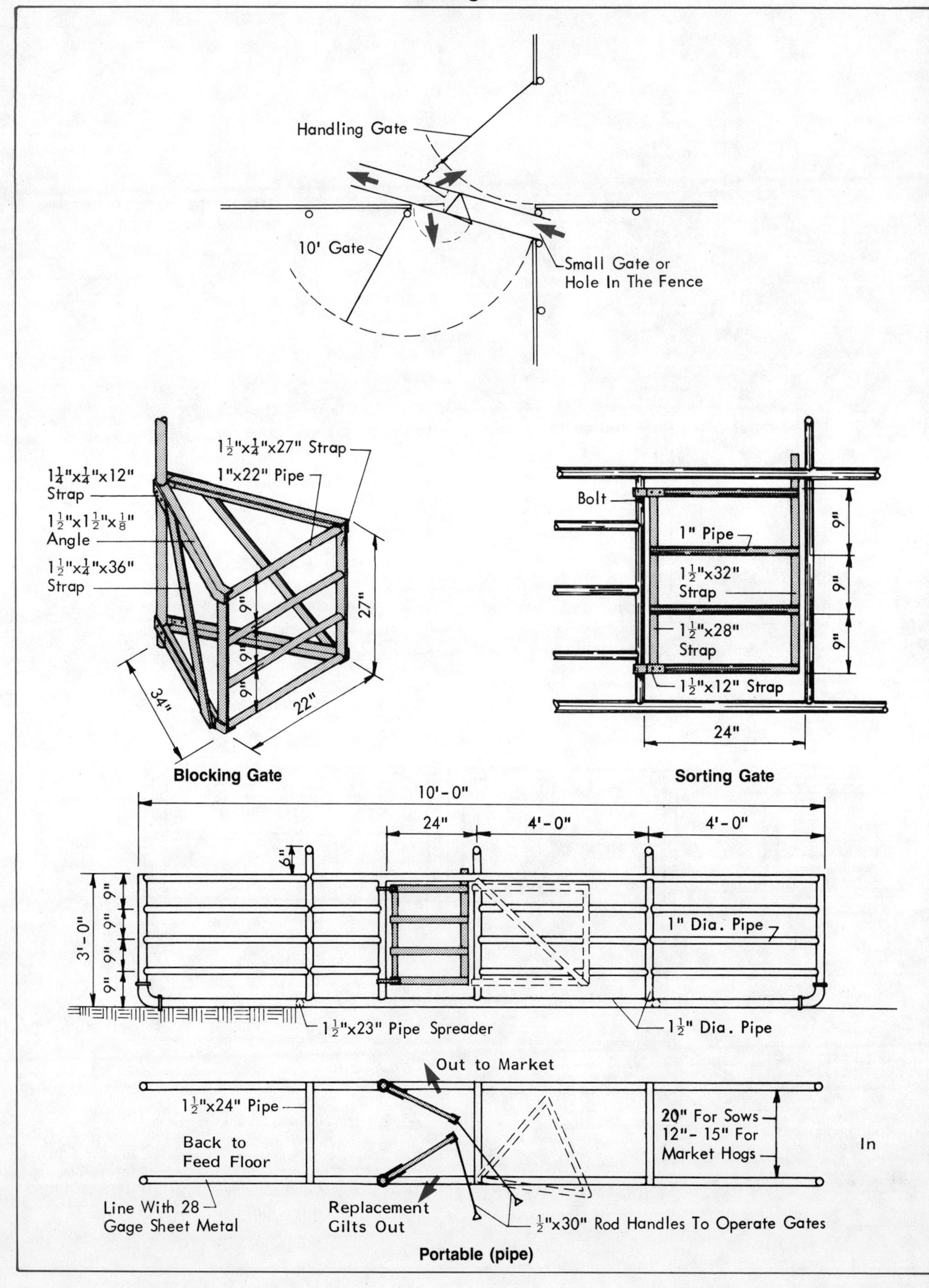

Breeding Rack

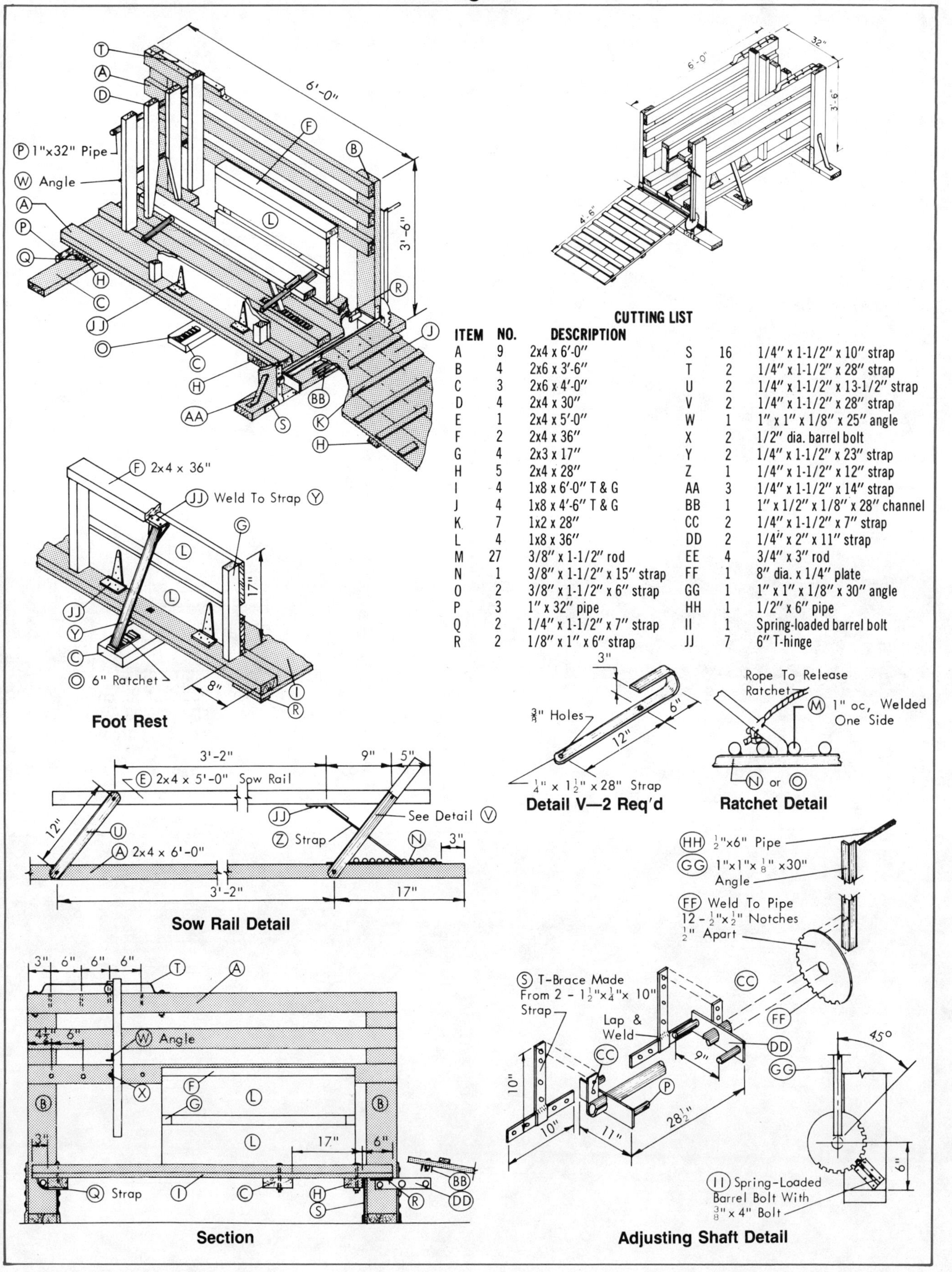

CUTTING LIST

ITEM	NO.	DESCRIPTION	ITEM	NO.	DESCRIPTION
A	9	2x4 x 6′-0″	S	16	1/4″ x 1-1/2″ x 10″ strap
B	4	2x6 x 3′-6″	T	2	1/4″ x 1-1/2″ x 28″ strap
C	3	2x6 x 4′-0″	U	2	1/4″ x 1-1/2″ x 13-1/2″ strap
D	4	2x4 x 30″	V	2	1/4″ x 1-1/2″ x 28″ strap
E	1	2x4 x 5′-0″	W	1	1″ x 1″ x 1/8″ x 25″ angle
F	2	2x4 x 36″	X	2	1/2″ dia. barrel bolt
G	4	2x3 x 17″	Y	2	1/4″ x 1-1/2″ x 23″ strap
H	5	2x4 x 28″	Z	1	1/4″ x 1-1/2″ x 12″ strap
I	4	1x8 x 6′-0″ T & G	AA	3	1/4″ x 1-1/2″ x 14″ strap
J	4	1x8 x 4′-6″ T & G	BB	1	1″ x 1/2″ x 1/8″ x 28″ channel
K	7	1x2 x 28″	CC	2	1/4″ x 1-1/2″ x 7″ strap
L	4	1x8 x 36″	DD	2	1/4″ x 2″ x 11″ strap
M	27	3/8″ x 1-1/2″ rod	EE	4	3/4″ x 3″ rod
N	1	3/8″ x 1-1/2″ x 15″ strap	FF	1	8″ dia. x 1/4″ plate
O	2	3/8″ x 1-1/2″ x 6″ strap	GG	1	1″ x 1″ x 1/8″ x 30″ angle
P	3	1″ x 32″ pipe	HH	1	1/2″ x 6″ pipe
Q	2	1/4″ x 1-1/2″ x 7″ strap	II	1	Spring-loaded barrel bolt
R	2	1/8″ x 1″ x 6″ strap	JJ	7	6″ T-hinge

Gestation Stalls

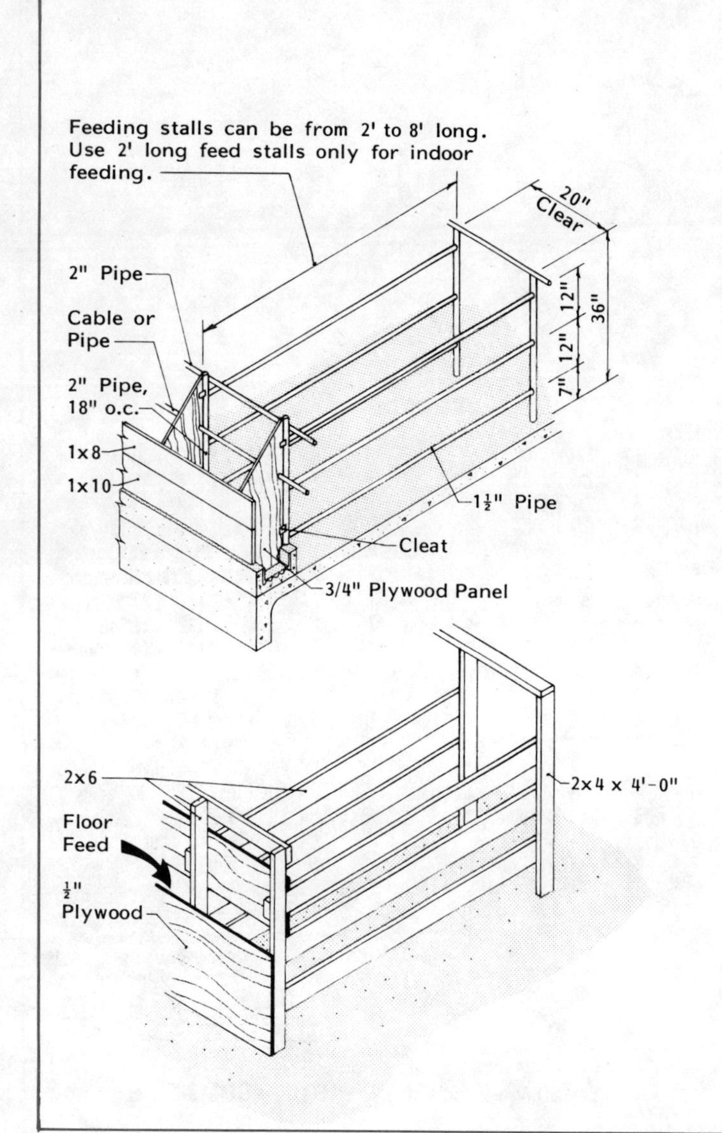

Set posts in 2½″ pipe sleeves or 4″ tile set in the concrete.

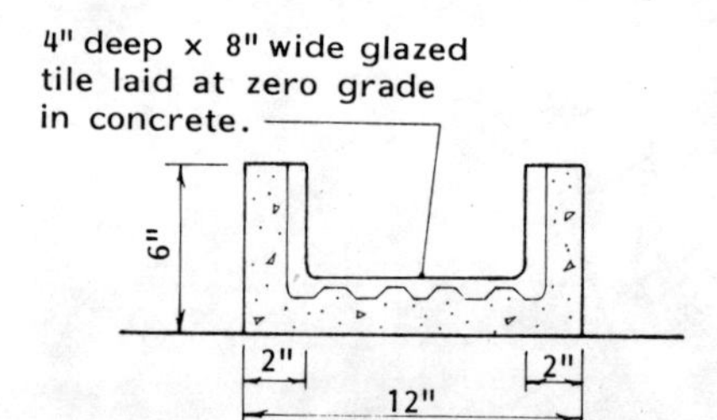

Sow Feed Trough Detail

Note: Provide a level trough, float controlled water delivery, and add feed to the water. Use trough with rounded corners for dry feed.

Farrowing Stalls

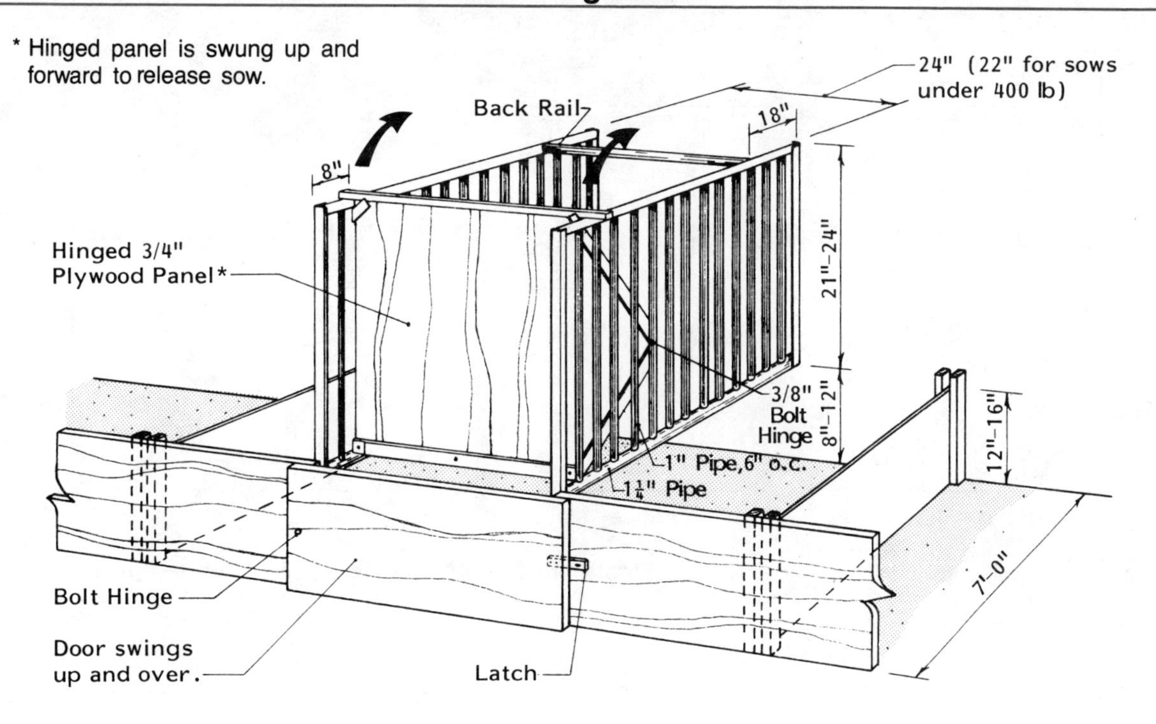

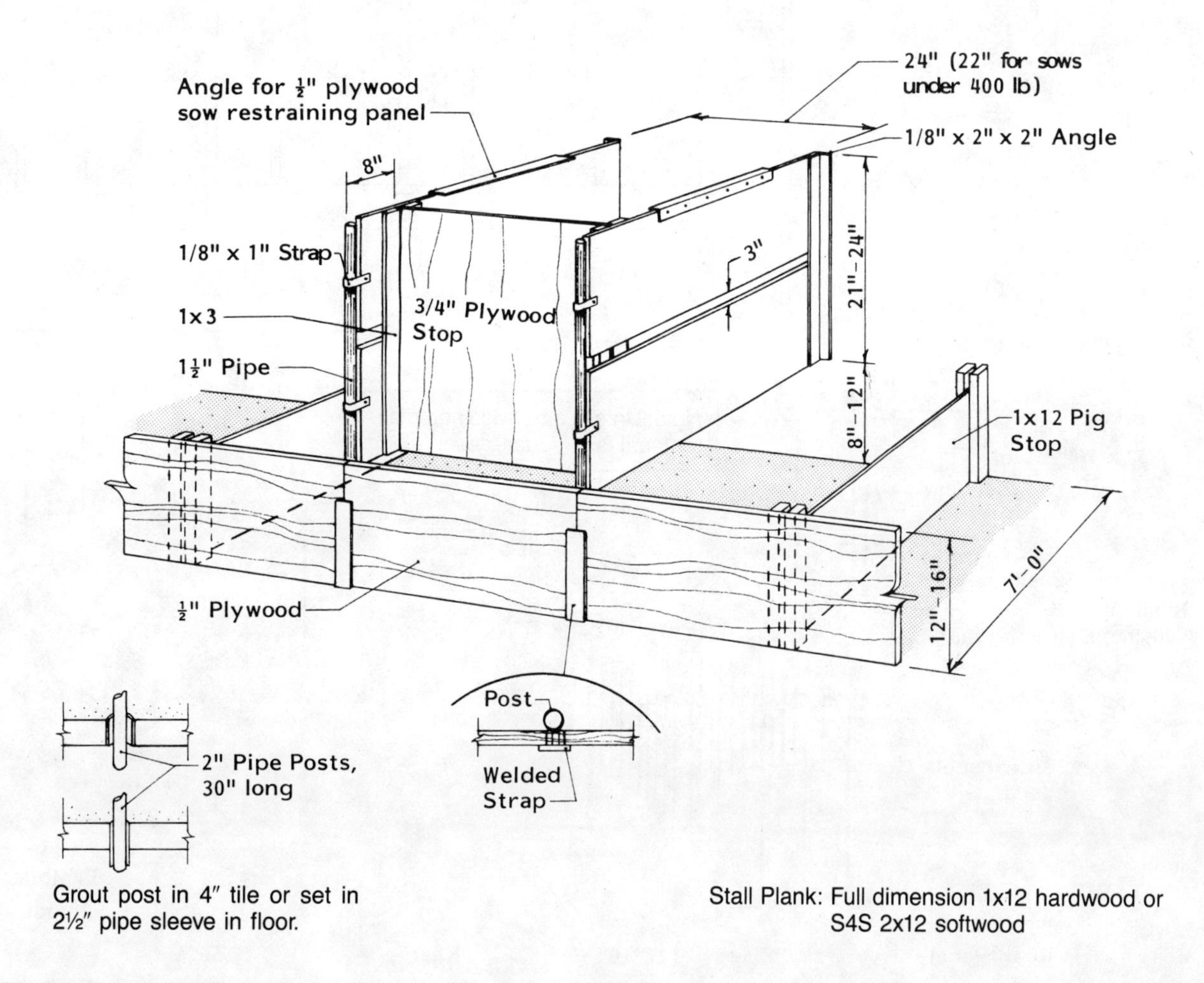

Grout post in 4″ tile or set in 2½″ pipe sleeve in floor.

Stall Plank: Full dimension 1x12 hardwood or S4S 2x12 softwood

Farrowing Stall

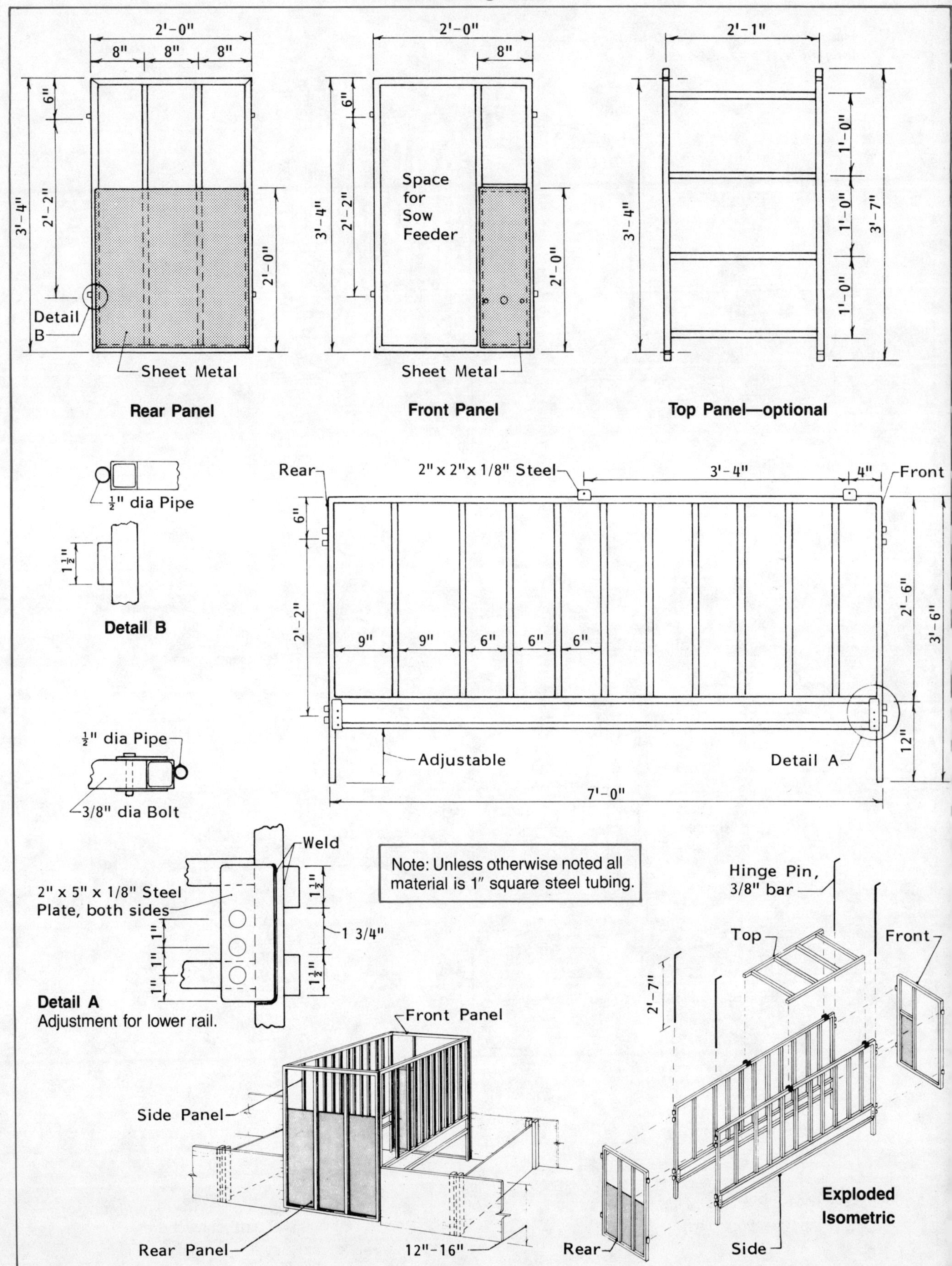

Farrowing Stall

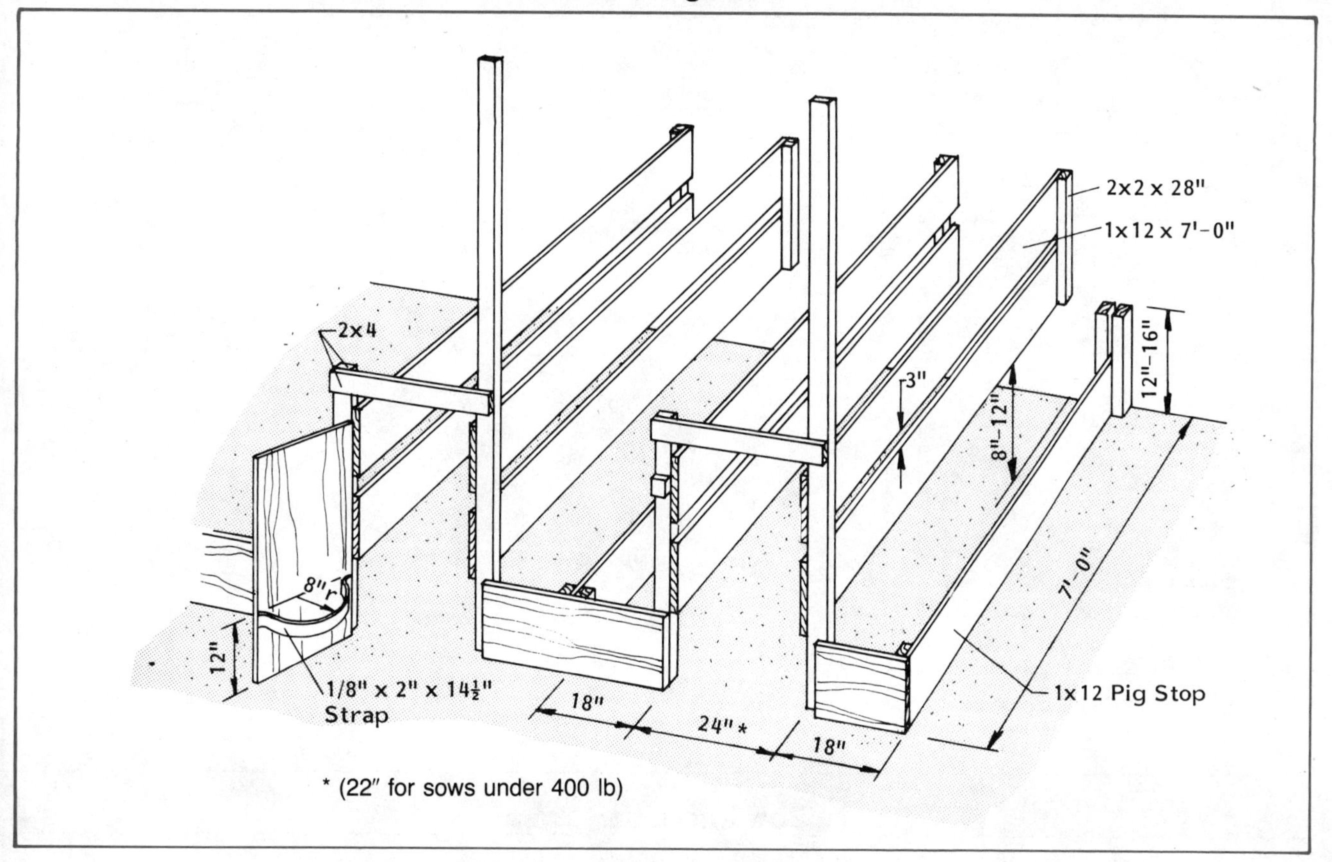

Farrowing—Growing Pen

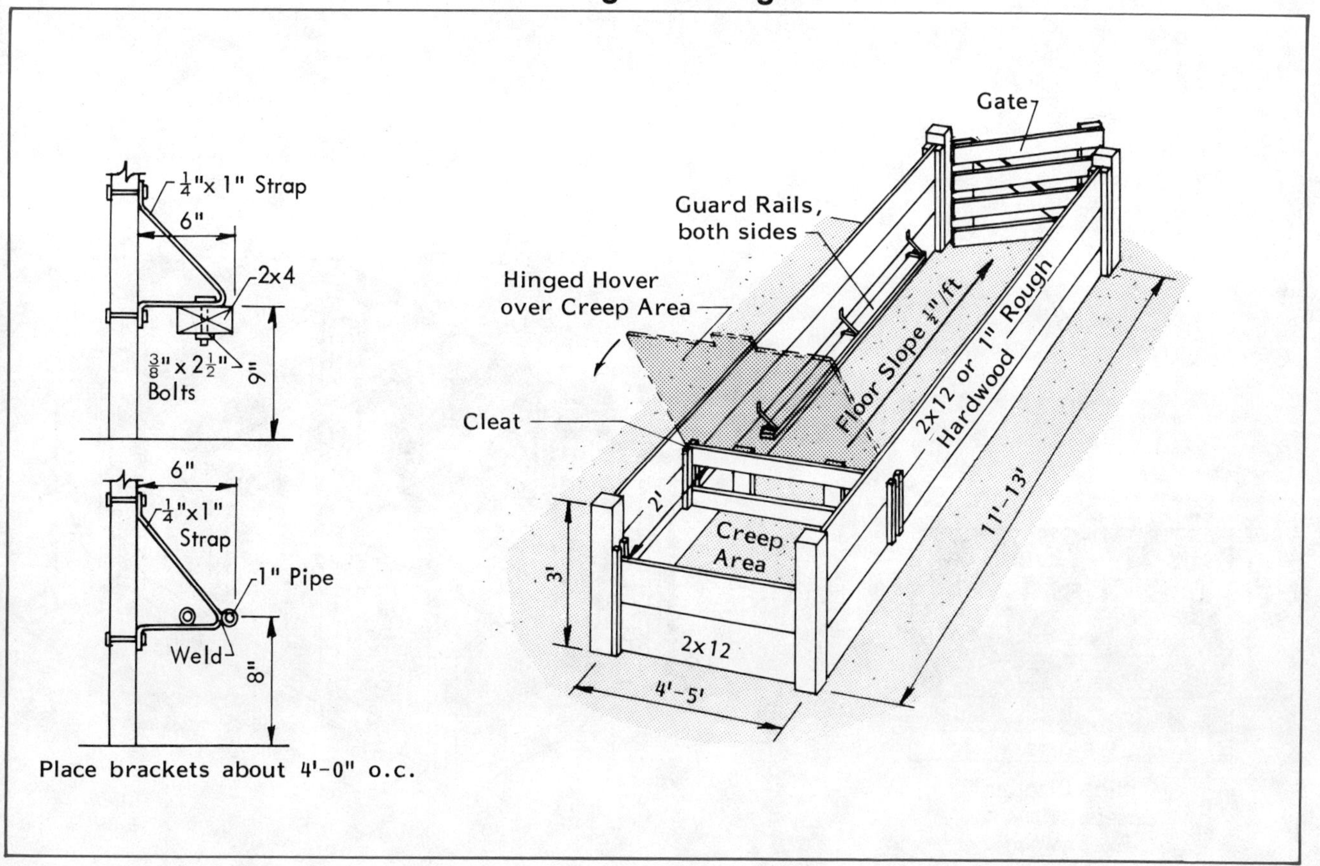

Farrowing Stall

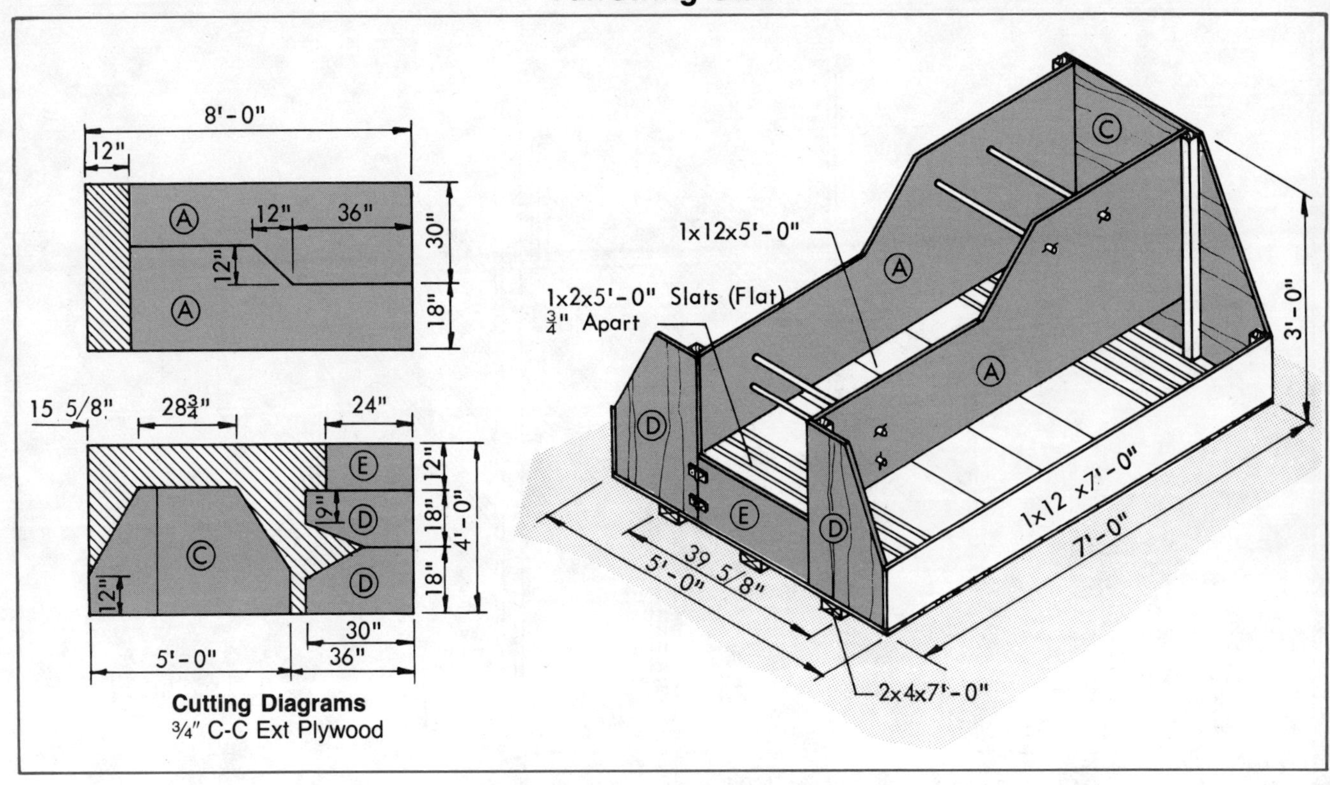

Sow and Litter Crate

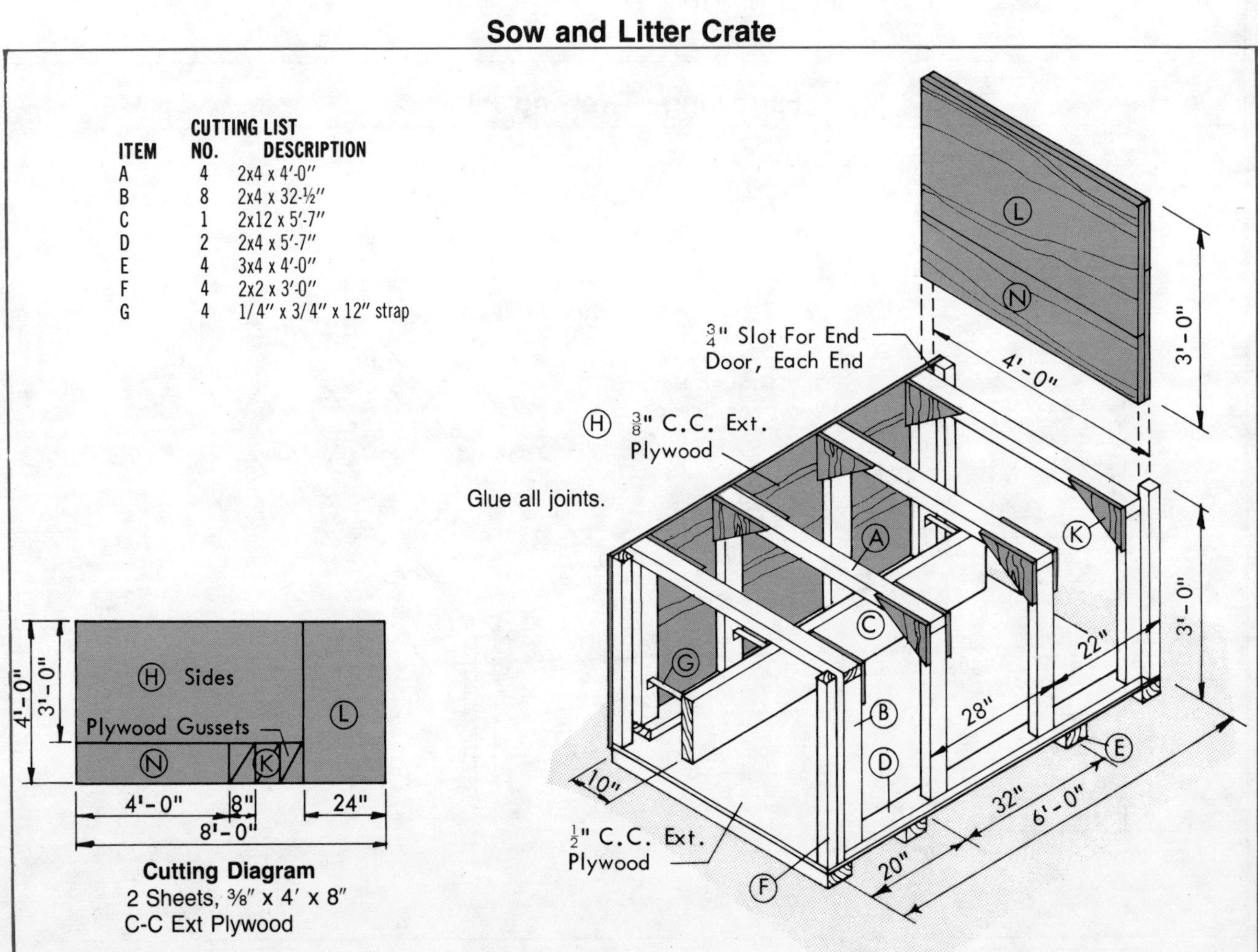

CUTTING LIST

ITEM	NO.	DESCRIPTION
A	4	2x4 x 4'-0"
B	8	2x4 x 32-½"
C	1	2x12 x 5'-7"
D	2	2x4 x 5'-7"
E	4	3x4 x 4'-0"
F	4	2x2 x 3'-0"
G	4	1/4" x 3/4" x 12" strap

Weighing Crate

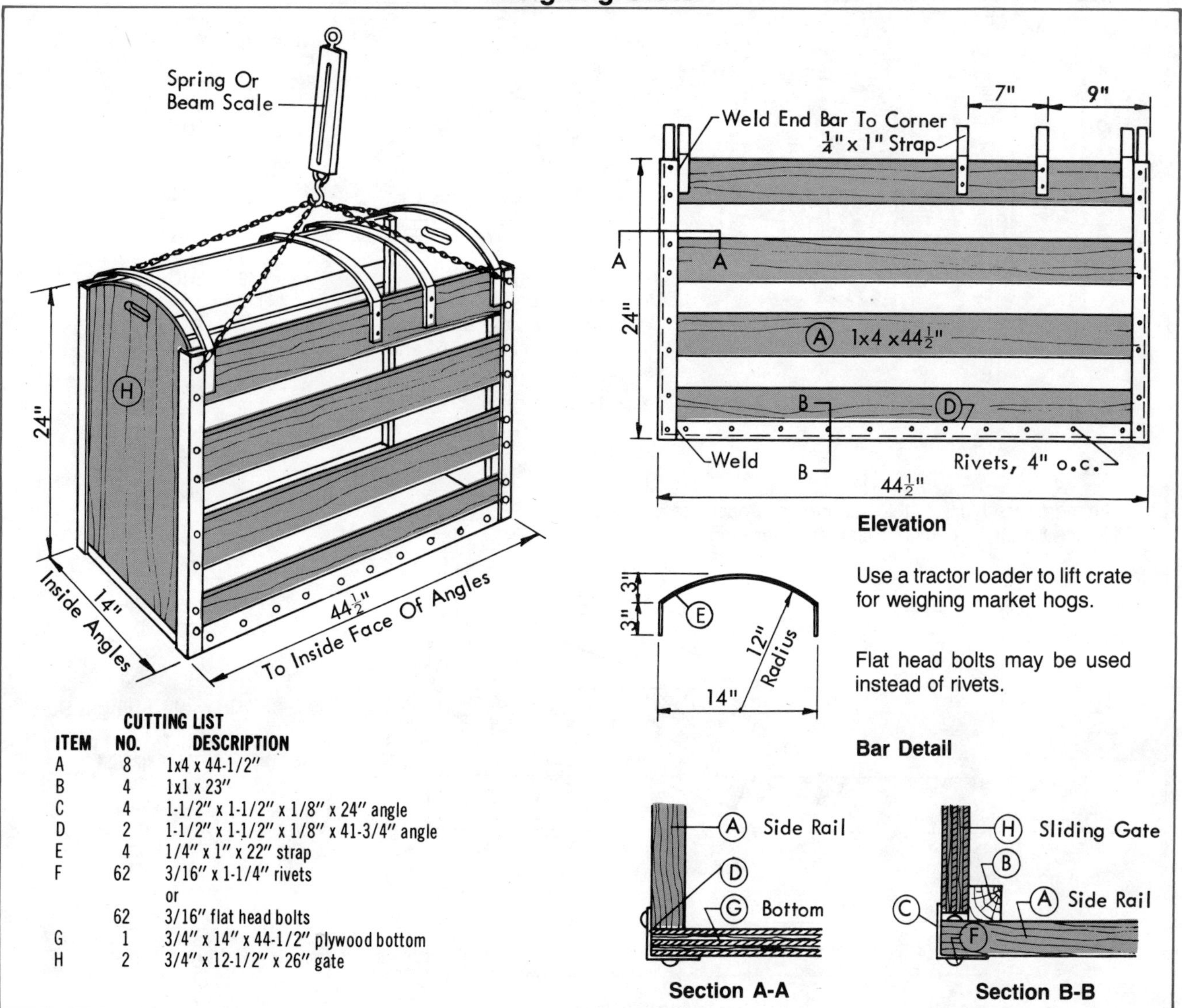

CUTTING LIST

ITEM	NO.	DESCRIPTION
A	8	1x4 x 44-1/2"
B	4	1x1 x 23"
C	4	1-1/2" x 1-1/2" x 1/8" x 24" angle
D	2	1-1/2" x 1-1/2" x 1/8" x 41-3/4" angle
E	4	1/4" x 1" x 22" strap
F	62	3/16" x 1-1/4" rivets
		or
	62	3/16" flat head bolts
G	1	3/4" x 14" x 44-1/2" plywood bottom
H	2	3/4" x 12-1/2" x 26" gate

Fork Lift

3-Point Hitch

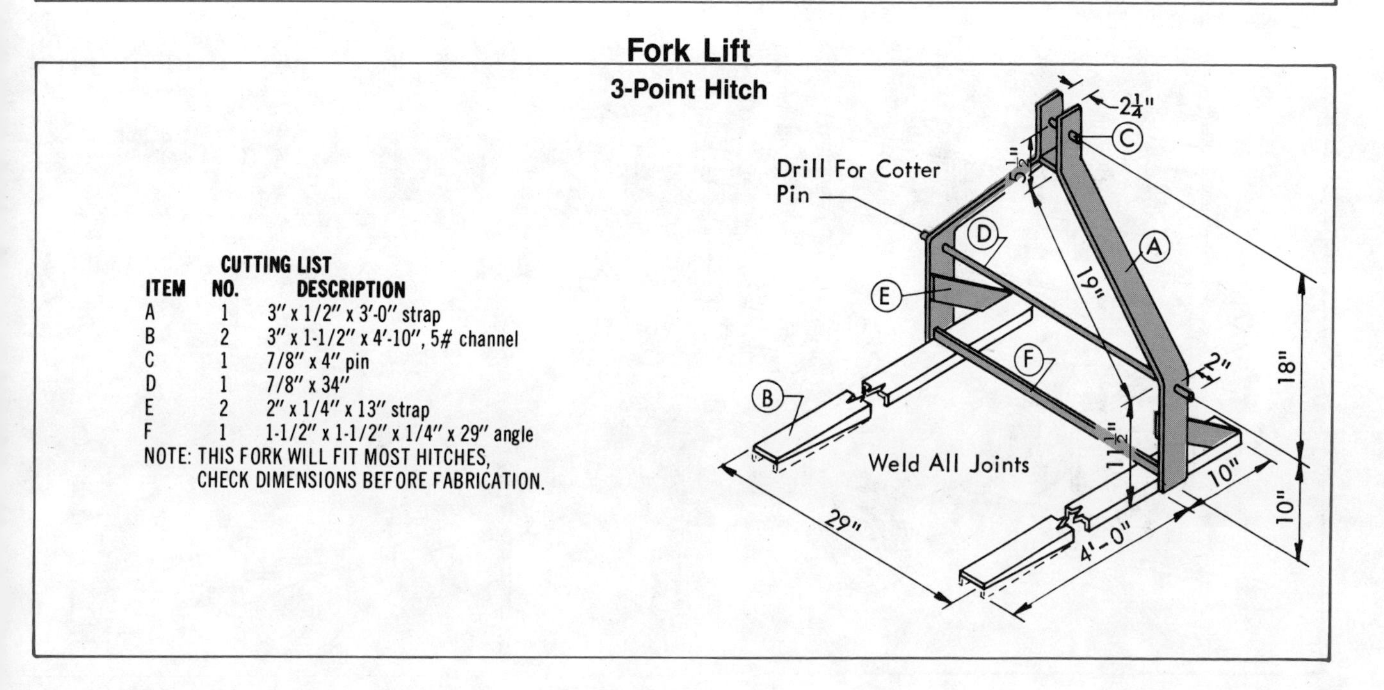

CUTTING LIST

ITEM	NO.	DESCRIPTION
A	1	3" x 1/2" x 3'-0" strap
B	2	3" x 1-1/2" x 4'-10", 5# channel
C	1	7/8" x 4" pin
D	1	7/8" x 34"
E	2	2" x 1/4" x 13" strap
F	1	1-1/2" x 1-1/2" x 1/4" x 29" angle

NOTE: THIS FORK WILL FIT MOST HITCHES, CHECK DIMENSIONS BEFORE FABRICATION.

Loading Chutes

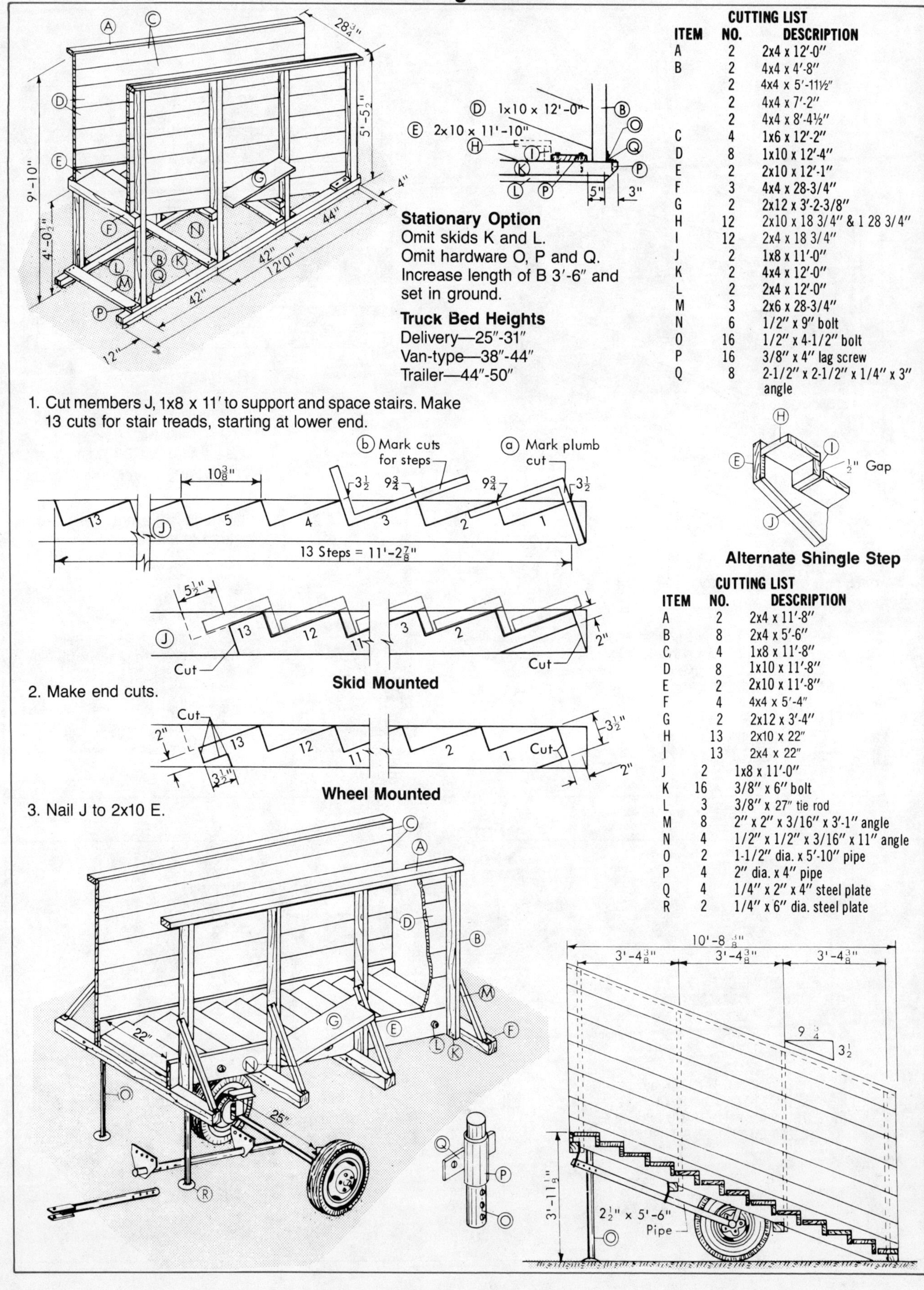

CUTTING LIST

ITEM	NO.	DESCRIPTION
A	2	2x4 x 12′-0″
B	2	4x4 x 4′-8″
	2	4x4 x 5′-11½″
	2	4x4 x 7′-2″
	2	4x4 x 8′-4½″
C	4	1x6 x 12′-2″
D	8	1x10 x 12′-4″
E	2	2x10 x 12′-1″
F	3	4x4 x 28-3/4″
G	2	2x12 x 3′-2-3/8″
H	12	2x10 x 18 3/4″ & 1 28 3/4″
I	12	2x4 x 18 3/4″
J	2	1x8 x 11′-0″
K	2	4x4 x 12′-0″
L	2	2x4 x 12′-0″
M	3	2x6 x 28-3/4″
N	6	1/2″ x 9″ bolt
O	16	1/2″ x 4-1/2″ bolt
P	16	3/8″ x 4″ lag screw
Q	8	2-1/2″ x 2-1/2″ x 1/4″ x 3″ angle

CUTTING LIST

ITEM	NO.	DESCRIPTION
A	2	2x4 x 11′-8″
B	8	2x4 x 5′-6″
C	4	1x8 x 11′-8″
D	8	1x10 x 11′-8″
E	2	2x10 x 11′-8″
F	4	4x4 x 5′-4″
G	2	2x12 x 3′-4″
H	13	2x10 x 22″
I	13	2x4 x 22″
J	2	1x8 x 11′-0″
K	16	3/8″ x 6″ bolt
L	3	3/8″ x 27″ tie rod
M	8	2″ x 2″ x 3/16″ x 3′-1″ angle
N	4	1/2″ x 1/2″ x 3/16″ x 11″ angle
O	2	1-1/2″ dia. x 5′-10″ pipe
P	4	2″ dia. x 4″ pipe
Q	4	1/4″ x 2″ x 4″ steel plate
R	2	1/4″ x 6″ dia. steel plate

Variable Height

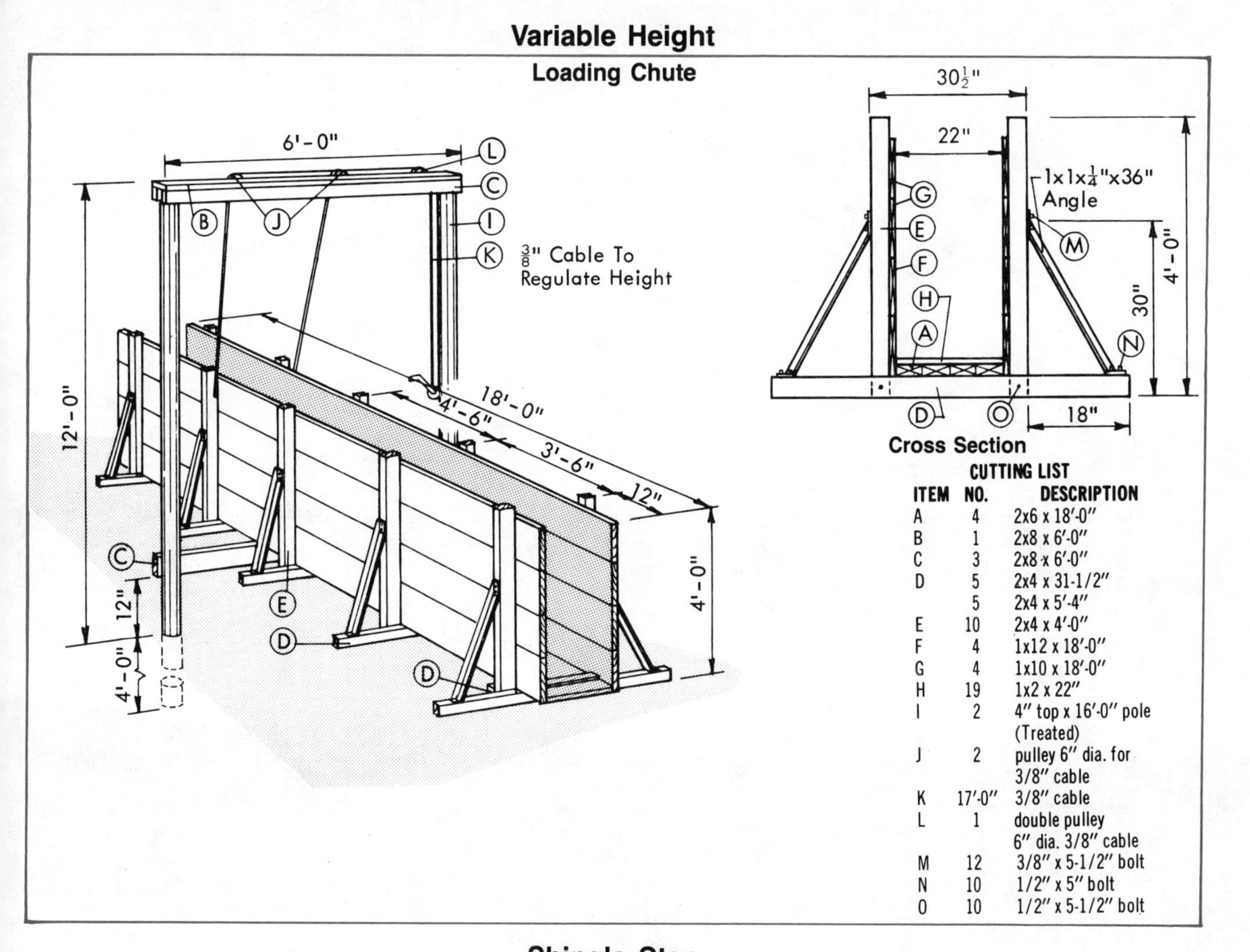

CUTTING LIST

ITEM	NO.	DESCRIPTION
A	4	2x6 x 18'-0"
B	1	2x8 x 6'-0"
C	3	2x8 x 6'-0"
D	5	2x4 x 31-1/2"
	5	2x4 x 5'-4"
E	10	2x4 x 4'-0"
F	4	1x12 x 18'-0"
G	4	1x10 x 18'-0"
H	19	1x2 x 22"
I	2	4" top x 16'-0" pole (Treated)
J	2	pulley 6" dia. for 3/8" cable
K	17'-0"	3/8" cable
L	1	double pulley 6" dia. 3/8" cable
M	12	3/8" x 5-1/2" bolt
N	10	1/2" x 5" bolt
O	10	1/2" x 5-1/2" bolt

Shingle Step

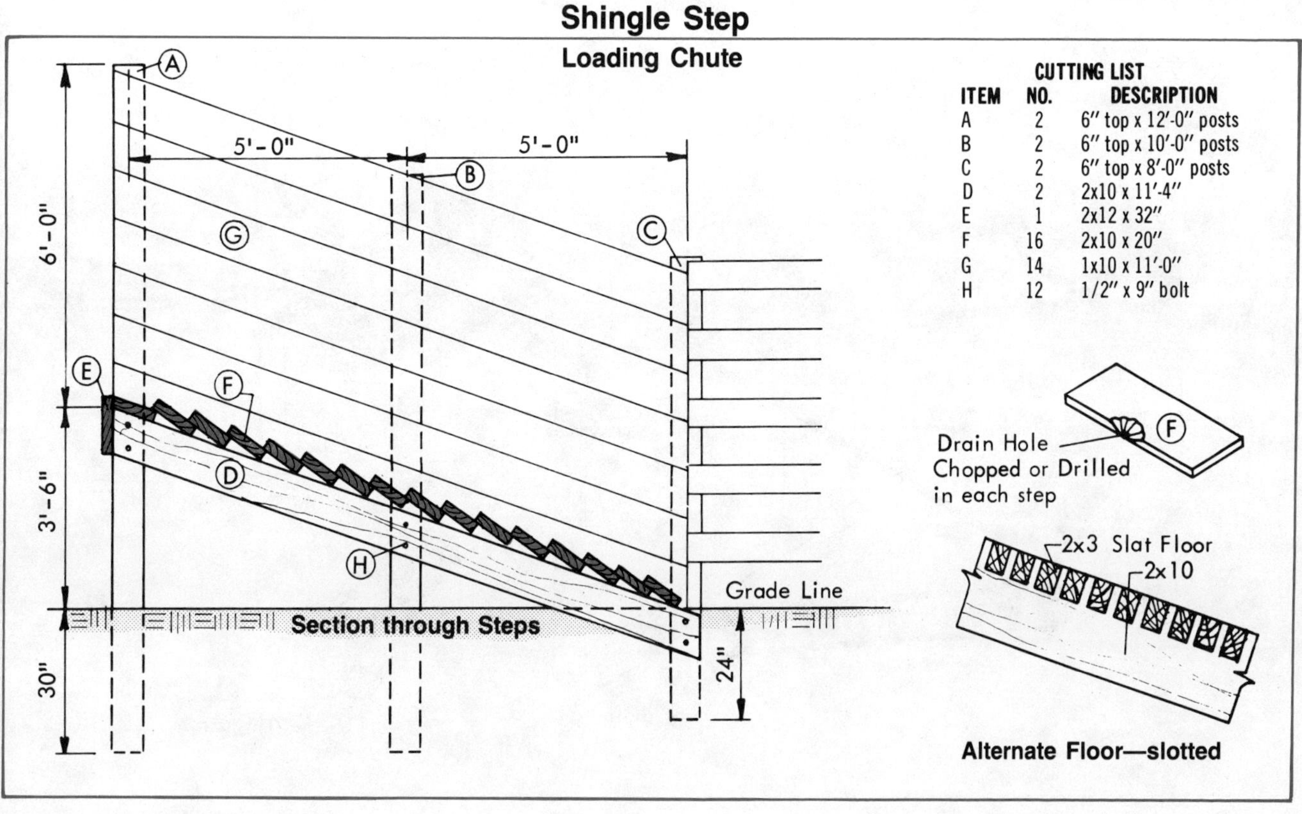

CUTTING LIST

ITEM	NO.	DESCRIPTION
A	2	6" top x 12'-0" posts
B	2	6" top x 10'-0" posts
C	2	6" top x 8'-0" posts
D	2	2x10 x 11'-4"
E	1	2x12 x 32"
F	16	2x10 x 20"
G	14	1x10 x 11'-0"
H	12	1/2" x 9" bolt

Portable Loading Chute

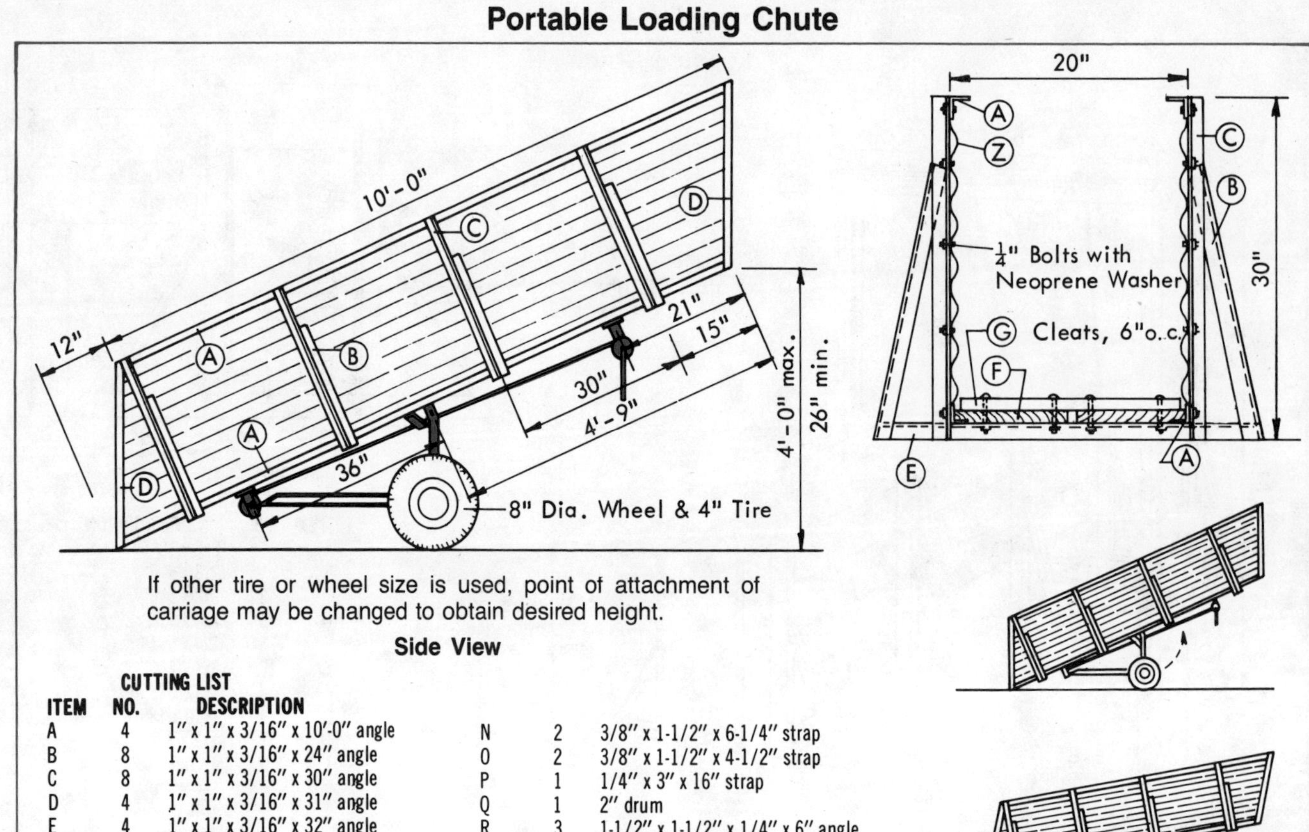

If other tire or wheel size is used, point of attachment of carriage may be changed to obtain desired height.

Side View

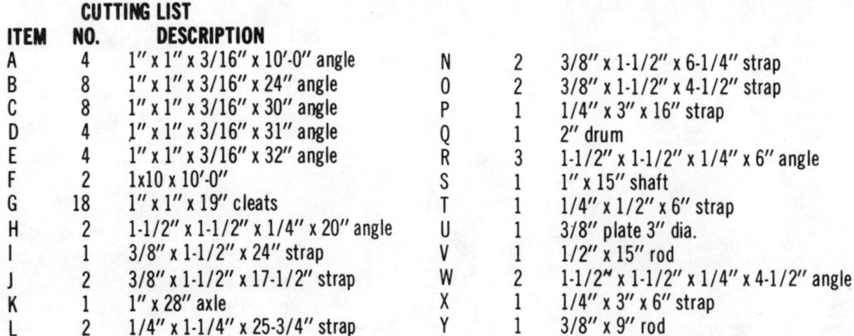

CUTTING LIST

ITEM	NO.	DESCRIPTION
A	4	1" x 1" x 3/16" x 10'-0" angle
B	8	1" x 1" x 3/16" x 24" angle
C	8	1" x 1" x 3/16" x 30" angle
D	4	1" x 1" x 3/16" x 31" angle
E	4	1" x 1" x 3/16" x 32" angle
F	2	1x10 x 10'-0"
G	18	1" x 1" x 19" cleats
H	2	1-1/2" x 1-1/2" x 1/4" x 20" angle
I	1	3/8" x 1-1/2" x 24" strap
J	2	3/8" x 1-1/2" x 17-1/2" strap
K	1	1" x 28" axle
L	2	1/4" x 1-1/4" x 25-3/4" strap
M	2	1-1/2" x 1-1/2" x 1/4" x 3" angle
N	2	3/8" x 1-1/2" x 6-1/4" strap
O	2	3/8" x 1-1/2" x 4-1/2" strap
P	1	1/4" x 3" x 16" strap
Q	1	2" drum
R	3	1-1/2" x 1-1/2" x 1/4" x 6" angle
S	1	1" x 15" shaft
T	1	1/4" x 1/2" x 6" strap
U	1	3/8" plate 3" dia.
V	1	1/2" x 15" rod
W	2	1-1/2" x 1-1/2" x 1/4" x 4-1/2" angle
X	1	1/4" x 3" x 6" strap
Y	1	3/8" x 9" rod
Z	2	28" x 12'-0" corr. metal

Weld all non-rotating parts. All bolts not specified are ¼". If other tire or wheel size is used, change length of members J and L to maintain desired length.

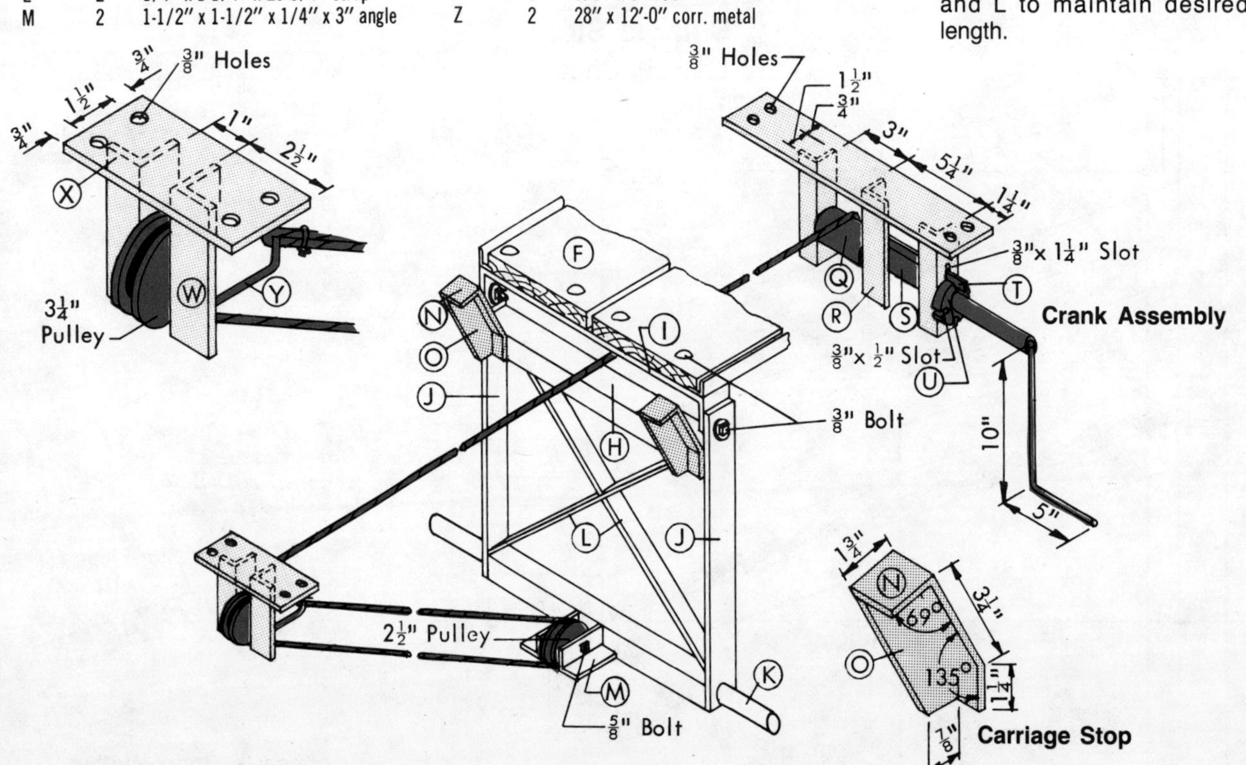

Trailer

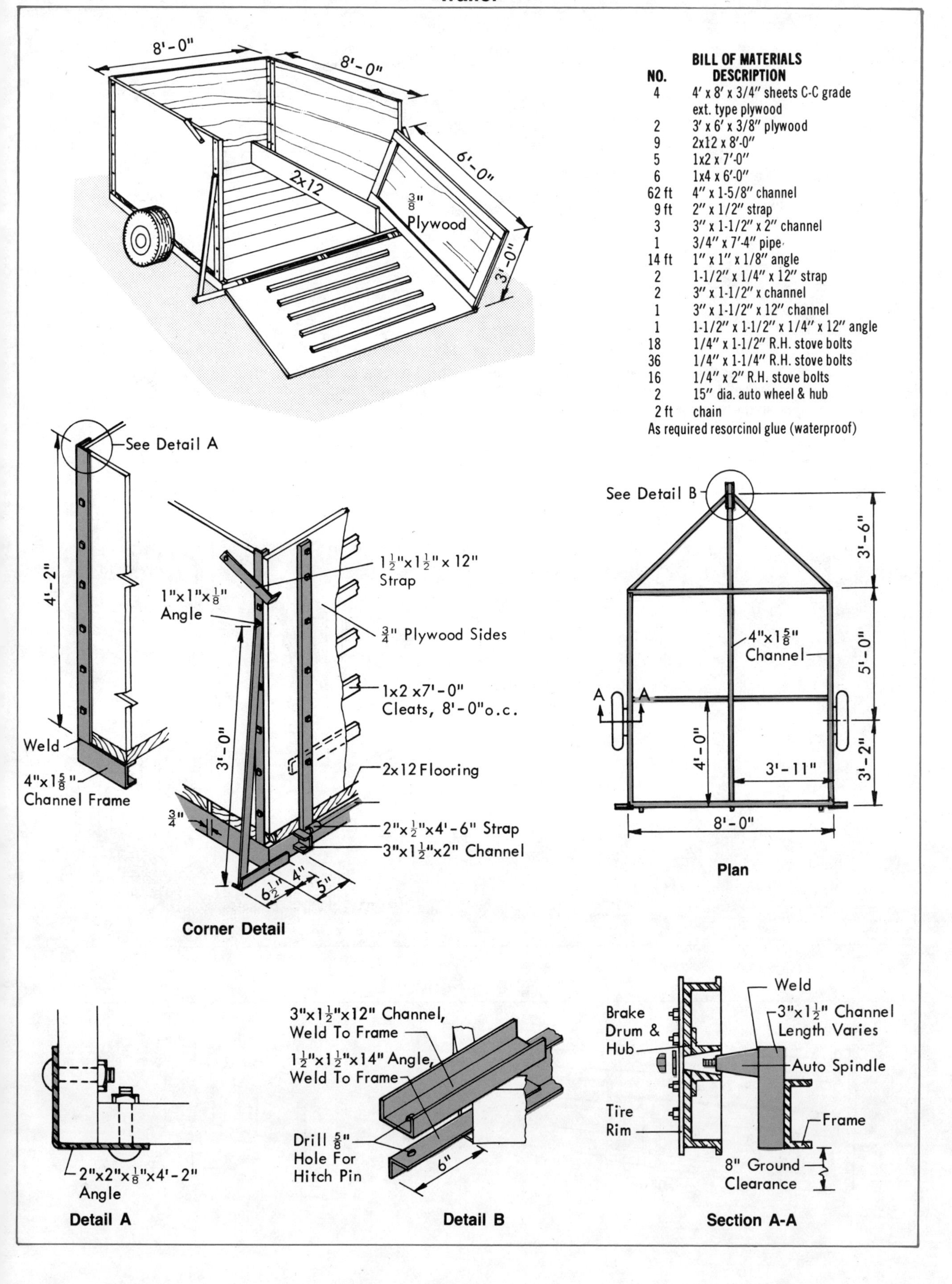

NO.	BILL OF MATERIALS DESCRIPTION
4	4′ x 8′ x 3/4″ sheets C-C grade ext. type plywood
2	3′ x 6′ x 3/8″ plywood
9	2x12 x 8′-0″
5	1x2 x 7′-0″
6	1x4 x 6′-0″
62 ft	4″ x 1-5/8″ channel
9 ft	2″ x 1/2″ strap
3	3″ x 1-1/2″ x 2″ channel
1	3/4″ x 7′-4″ pipe
14 ft	1″ x 1″ x 1/8″ angle
2	1-1/2″ x 1/4″ x 12″ strap
2	3″ x 1-1/2″ x channel
1	3″ x 1-1/2″ x 12″ channel
1	1-1/2″ x 1-1/2″ x 1/4″ x 12″ angle
18	1/4″ x 1-1/2″ R.H. stove bolts
36	1/4″ x 1-1/4″ R.H. stove bolts
16	1/4″ x 2″ R.H. stove bolts
2	15″ dia. auto wheel & hub
2 ft	chain

As required resorcinol glue (waterproof)

Corner Detail

Plan

Detail A

Detail B

Section A-A

FEEDING

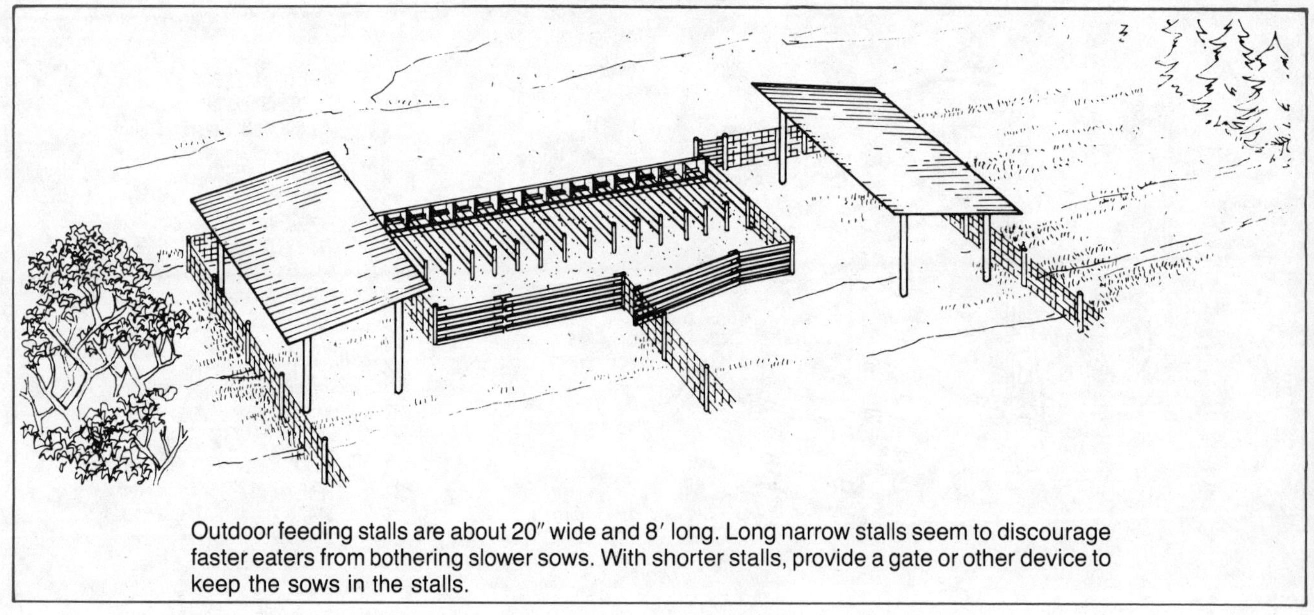

Outdoor feeding stalls are about 20″ wide and 8′ long. Long narrow stalls seem to discourage faster eaters from bothering slower sows. With shorter stalls, provide a gate or other device to keep the sows in the stalls.

Fenceline Feeders

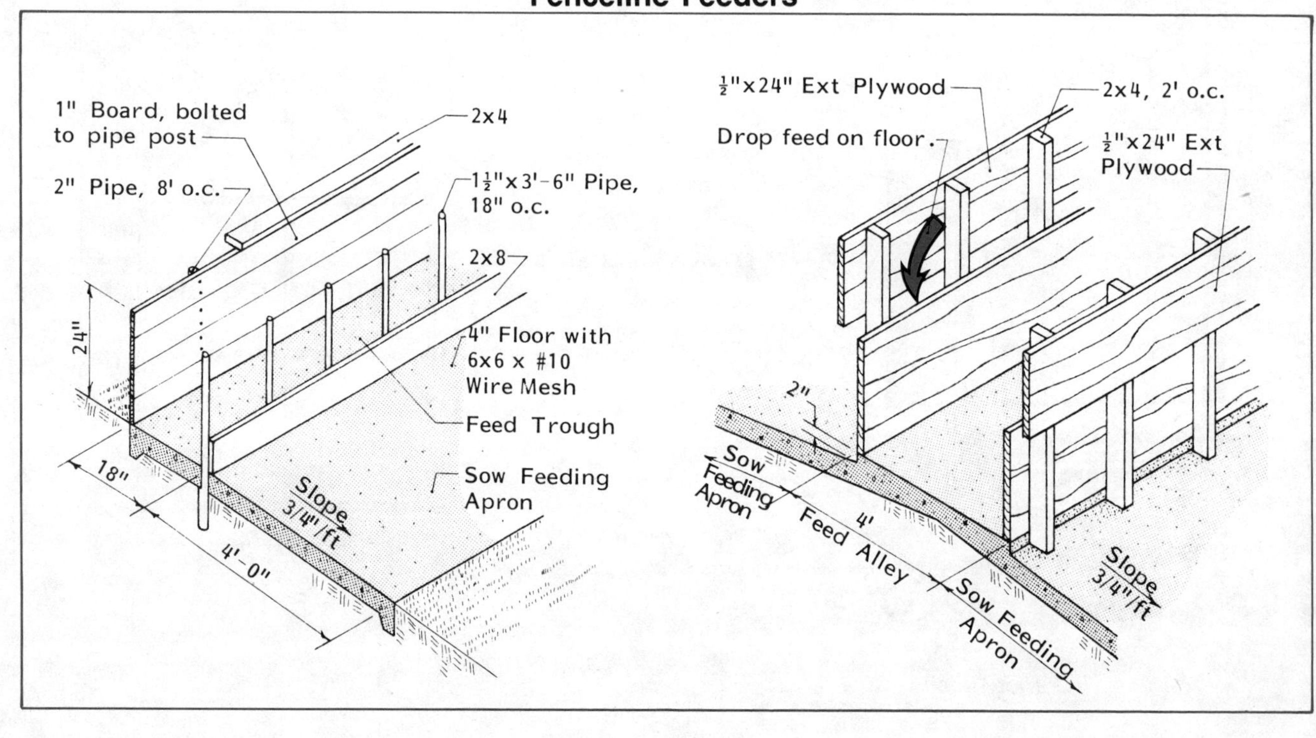

Creep Fence

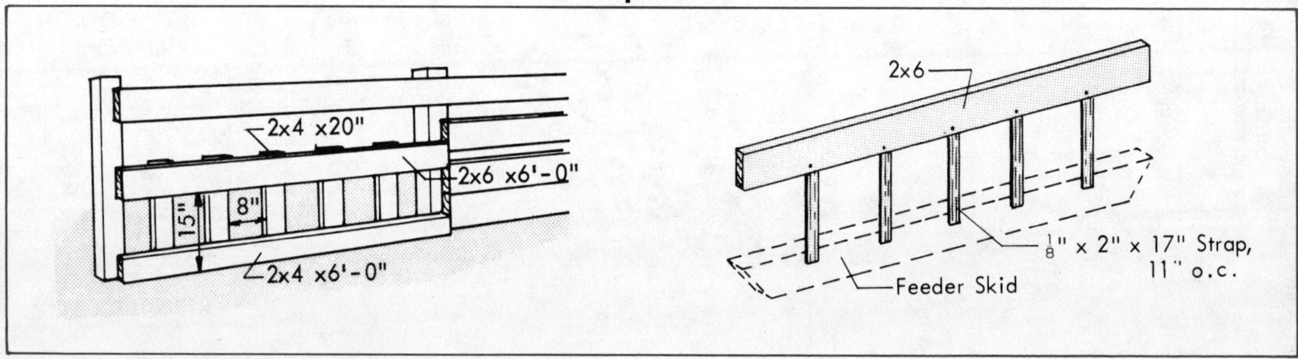

Finishing Feeders

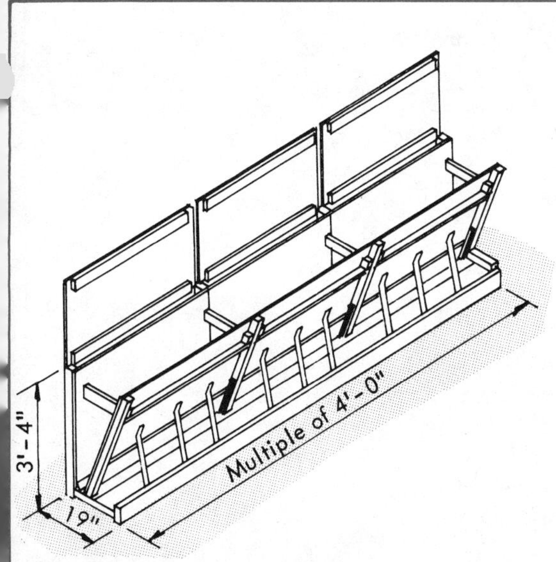

Capacity: 8 Bu per 4′ length
Locate along building wall for confined finishing, or along drive or fenceline. Provide roof for outdoor use.

Reduce capacity for wet corn: Suspend auger below J. Capacity 5 Bu per 4′.

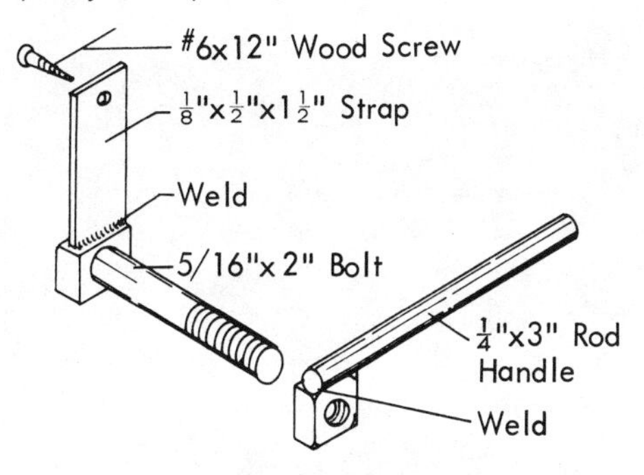

Assembly O Detail

CUTTING LIST

ITEM	NUMBER FIRST SECTION	NUMBER ADD. SECTION	DESCRIPTION
A	2	1	2x3 x 3′-4″
B	2	1	2x3 x 3′-6″
C	4	4* - 46-3/8″	1x3 x 3′-9-1/2″
D	2	2	1x2 x 15″
E	1	1* - 46-3/8″	1x10 x 3′-9-1/2″
F	2	2	2x8 x (Note)
G	1	1	2x6 x (Note)
H	2	2	2x2 x 17-5/8″
I	2	3	1x4 x 5″
J	2	1	1x4 x 22″
K,L,M	1 sht.	1 sht.	4′-8′ x 3/8″ C-C, ext. plywood
N	1	1	2x2 x 24″
O	2	2	assemblies
P	3	3	1/4″ x 3/4″ x 18″ strap
Q	3	3	5/16″ x 1″ bolt
R	3	3	5/16″ x 2″ bolt
S	4	2	3/8″ x 3-1/2″ bolt
T	3	3	3/8″ x 4″ lag screw
U	1	1	6″ T-hinge
V	1	0	6″ strap hinge

*Last section, 45-1/2″. Close ends.
Note: Use 12′ to 20′ boards for members F & G. (4′-0″ per section)

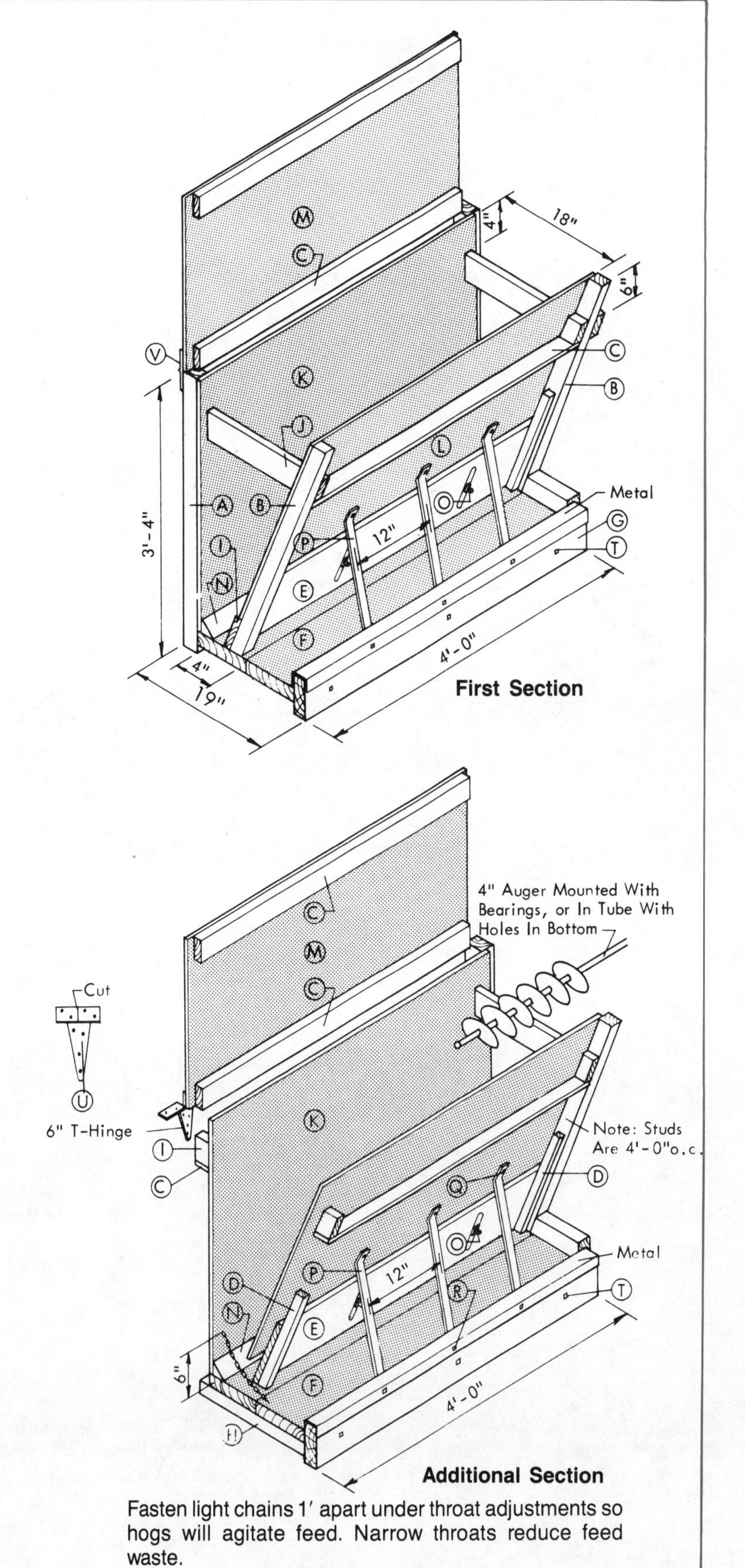

Additional Section

Fasten light chains 1′ apart under throat adjustments so hogs will agitate feed. Narrow throats reduce feed waste.

Finish Feeder

Capacity: 13 Bu per 4′ length
32 Pigs per 4′ length

Locate along pen partitions, or fencelines, or a sloping concrete apron. Provide roof for outdoor use.

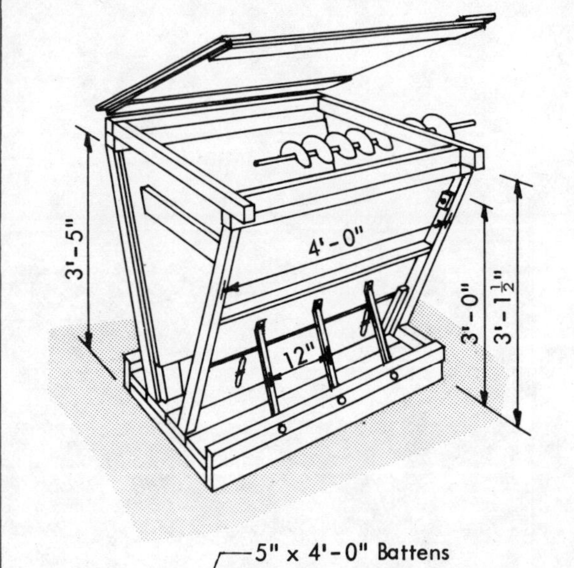

CUTTING LIST

ITEM	1ST SECTION DESCRIPTION	NO.	ADDITIONAL SECTION NO.
A	2x3 x 4′-0″	2	1
B	2x3 x 39-1/2″	2	2
C	2x3 x 42-1/2″	2	1
D	2x3 x 46-3/8″	2	2
E	2x8 x 49-5/8″	4 (Note)	4 - 48″
F	2x6 x 49-5/8″	2 (Note)	2 - 48″
G	1 x 4 x 3′-6″	2	1
H	1 x 3 x 46-3/8″	2	2
I	3/8″ x 5″ x 4′-0″ (plywood)	5	0
K	2x3 x 49-5/8″	2 (Note)	2
L	3′-3″ x 4′-0″ x 3/8″ (plywood)	1	1
M	2′-9″ x 4′-0″ x 3/8″ (plywood)	1	1
N	1x4 x 5″	4	4
O	1x10 x 46-3/8″	2	2
P	1x2 x 15″	4	4
Q	1/4″ x 3/4″ x 16″ metal bar	6	6
R	5/16″ bolt	6	6
S	3/8″ bolt	4	2
T	1-1/2″ x 24-gage x 5″ strap	4	2
U	5-5/8″ x 24-gage x 49-5/8″	2	2
V	4′ x 4′ x 3/8″ (plywood)	1	1
W	Provide plywood for ends	2	0
X	2x2 x 30″	2	2

Note: Use 12′ to 20′ boards for members E, F, & K. (4′-0″ per section)

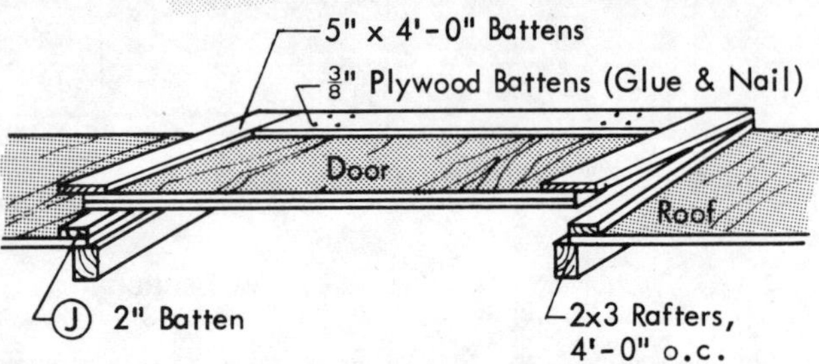

Roof Door Section

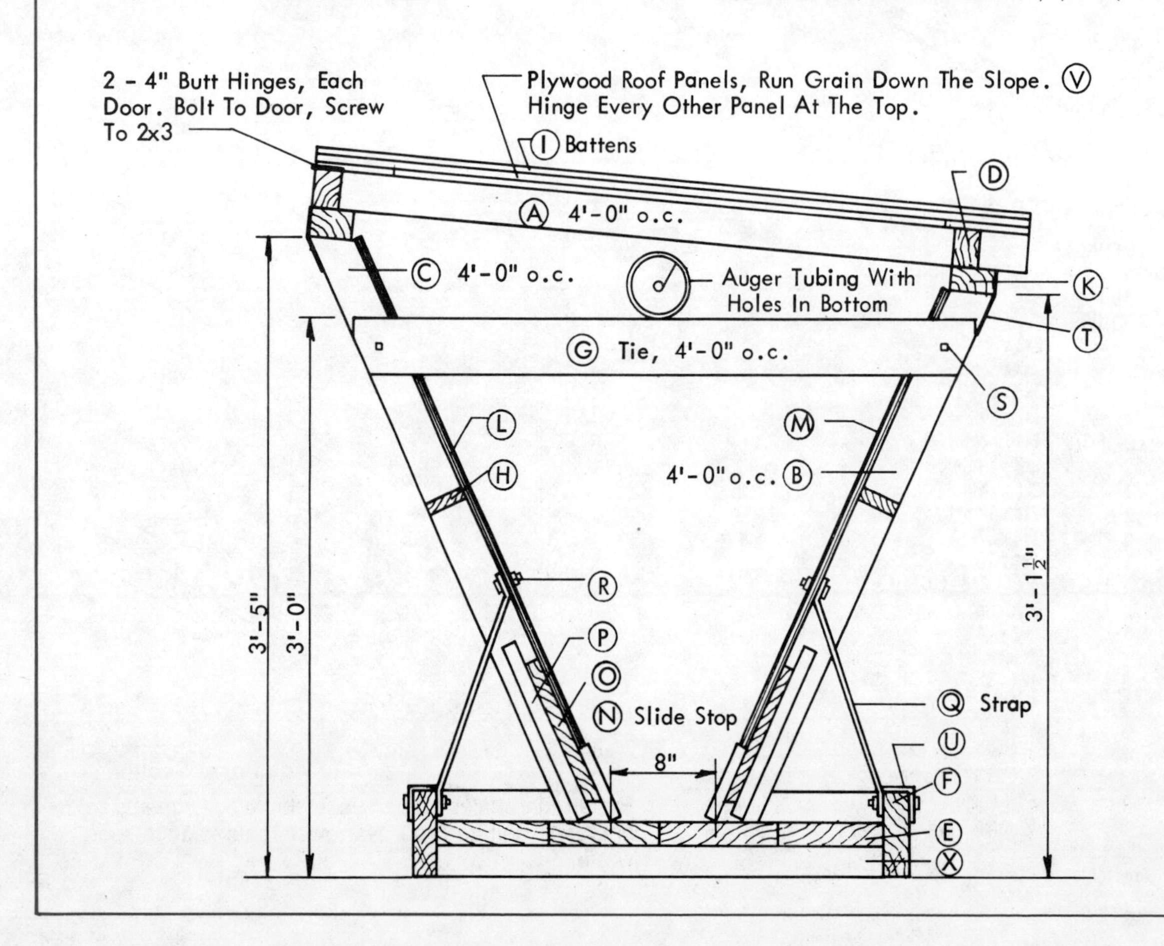

Walk-in Self-feeder

Plan A, 150 Bu—64 Pigs

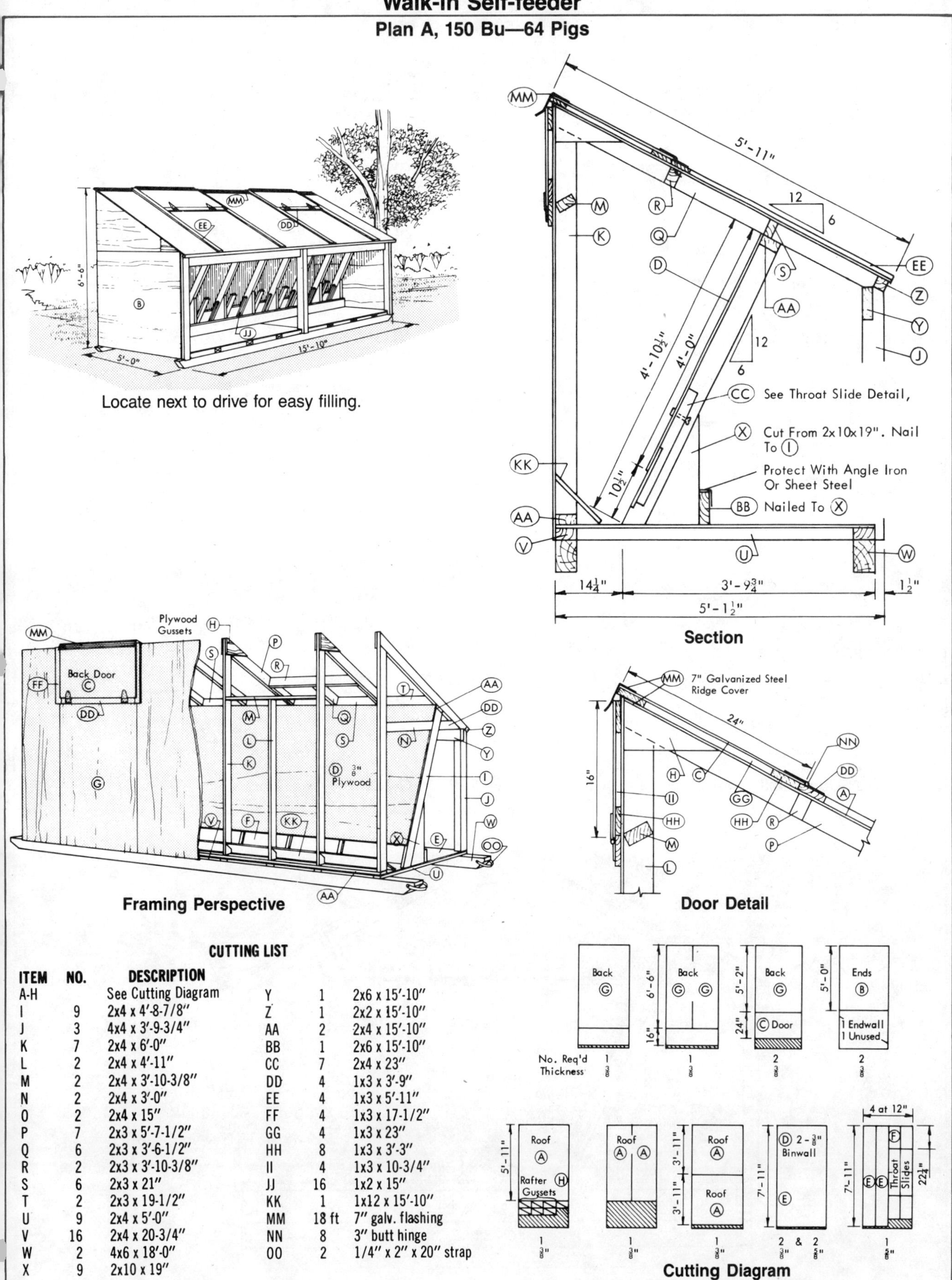

CUTTING LIST

ITEM	NO.	DESCRIPTION	ITEM	NO.	DESCRIPTION
A-H		See Cutting Diagram	Y	1	2x6 x 15'-10"
I	9	2x4 x 4'-8-7/8"	Z	1	2x2 x 15'-10"
J	3	4x4 x 3'-9-3/4"	AA	2	2x4 x 15'-10"
K	7	2x4 x 6'-0"	BB	1	2x6 x 15'-10"
L	2	2x4 x 4'-11"	CC	7	2x4 x 23"
M	2	2x4 x 3'-10-3/8"	DD	4	1x3 x 3'-9"
N	2	2x4 x 3'-0"	EE	4	1x3 x 5'-11"
O	2	2x4 x 15"	FF	4	1x3 x 17-1/2"
P	7	2x3 x 5'-7-1/2"	GG	4	1x3 x 23"
Q	6	2x3 x 3'-6-1/2"	HH	8	1x3 x 3'-3"
R	2	2x3 x 3'-10-3/8"	II	4	1x3 x 10-3/4"
S	6	2x3 x 21"	JJ	16	1x2 x 15"
T	2	2x3 x 19-1/2"	KK	1	1x12 x 15'-10"
U	9	2x4 x 5'-0"	MM	18 ft	7" galv. flashing
V	16	2x4 x 20-3/4"	NN	8	3" butt hinge
W	2	4x6 x 18'-0"	OO	2	1/4" x 2" x 20" strap
X	9	2x10 x 19"			

Walk-in Self-feeder

Plan B, 150 Bu—64 Pigs

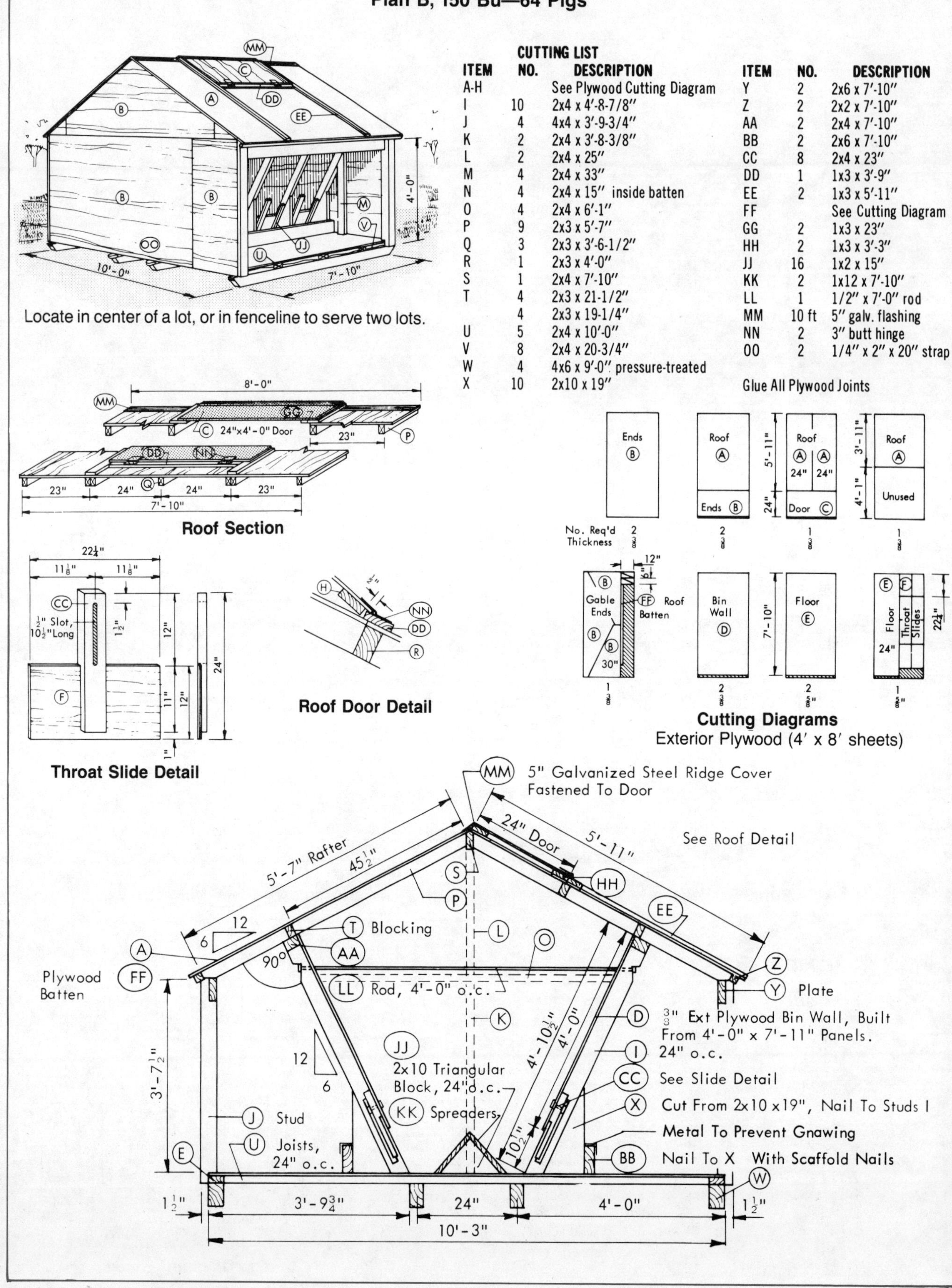

Locate in center of a lot, or in fenceline to serve two lots.

CUTTING LIST

ITEM	NO.	DESCRIPTION
A-H		See Plywood Cutting Diagram
I	10	2x4 x 4'-8-7/8"
J	4	4x4 x 3'-9-3/4"
K	2	2x4 x 3'-8-3/8"
L	2	2x4 x 25"
M	4	2x4 x 33"
N	4	2x4 x 15" inside batten
O	4	2x4 x 6'-1"
P	9	2x3 x 5'-7"
Q	3	2x3 x 3'-6-1/2"
R	1	2x3 x 4'-0"
S	1	2x4 x 7'-10"
T	4	2x3 x 21-1/2"
	4	2x3 x 19-1/4"
U	5	2x4 x 10'-0"
V	8	2x4 x 20-3/4"
W	4	4x6 x 9'-0" pressure-treated
X	10	2x10 x 19"
Y	2	2x6 x 7'-10"
Z	2	2x2 x 7'-10"
AA	2	2x4 x 7'-10"
BB	2	2x6 x 7'-10"
CC	8	2x4 x 23"
DD	1	1x3 x 3'-9"
EE	2	1x3 x 5'-11"
FF		See Cutting Diagram
GG	2	1x3 x 23"
HH	2	1x3 x 3'-3"
JJ	16	1x2 x 15"
KK	2	1x12 x 7'-10"
LL	1	1/2" x 7'-0" rod
MM	10 ft	5" galv. flashing
NN	2	3" butt hinge
OO	2	1/4" x 2" x 20" strap

Glue All Plywood Joints

Roof Section

Throat Slide Detail

Roof Door Detail

Cutting Diagrams

Exterior Plywood (4' x 8' sheets)

Feed Cart

10 Bu

16g x 3" Wide Sheet Metal

Corner Detail

Alternate

1/2" C-C Ext Plywood

18" 18"

14" 10" 10" 14"

Side | End | End

Side | Bottom

24"

12" 32" 12" 33"

25"

$25\frac{1}{2}$"

Mount Wheels & Casters On 2" Plank

4'-8"

Concrete Trough

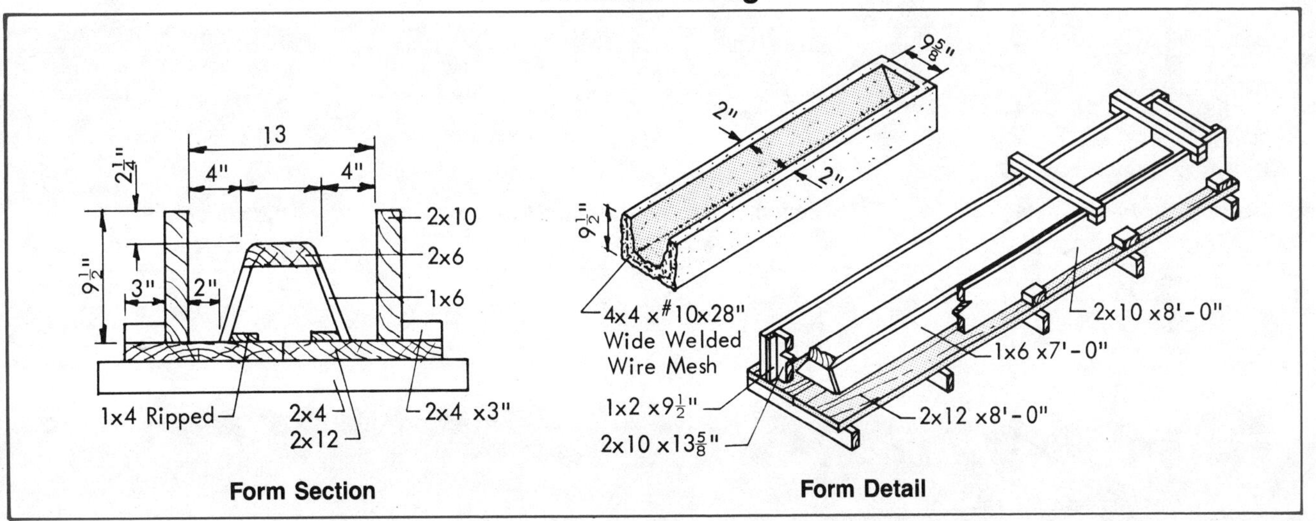

Form Section

Form Detail

Auger Installations

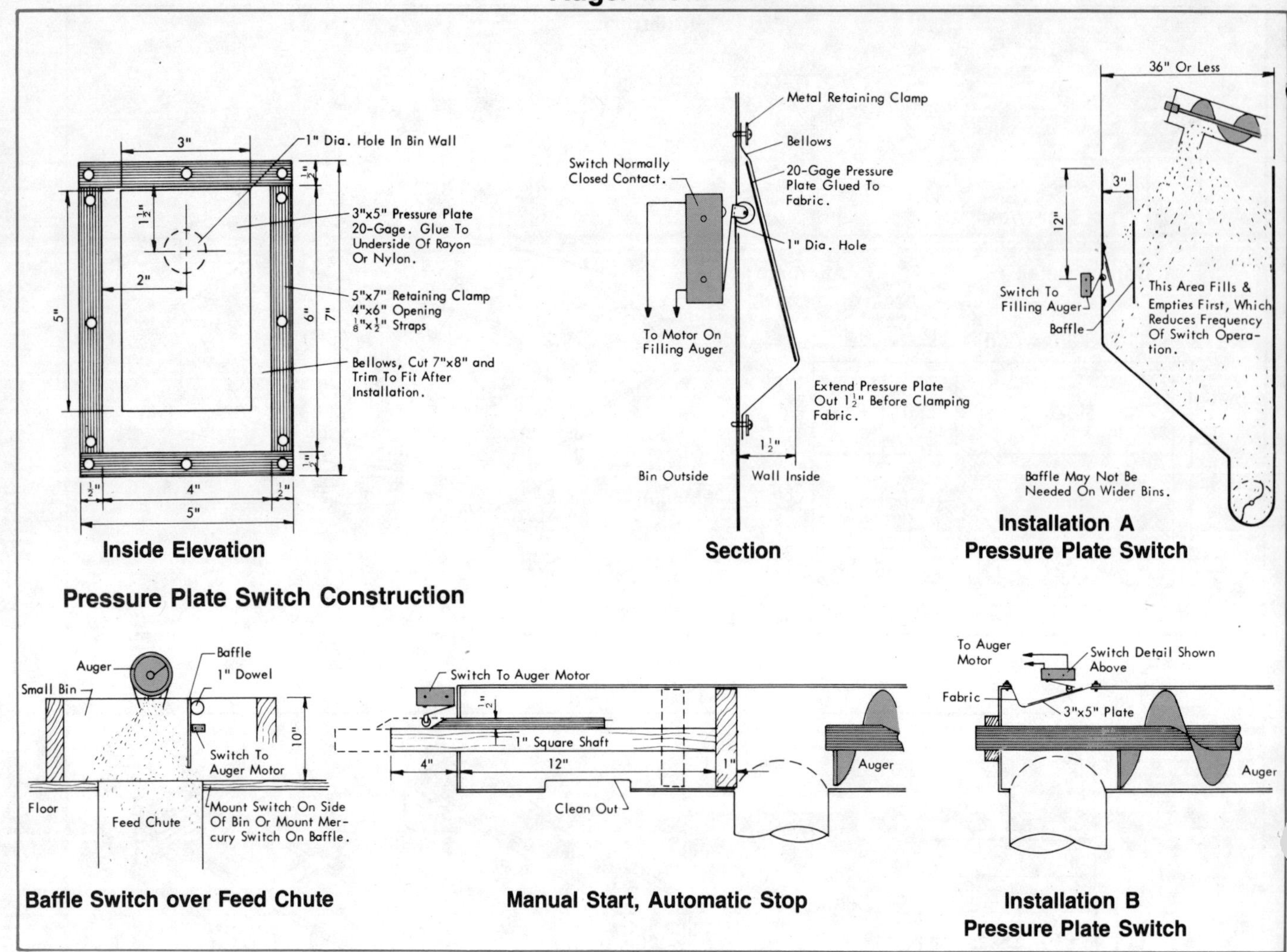

Inside Elevation

Section

Installation A
Pressure Plate Switch

Pressure Plate Switch Construction

Baffle Switch over Feed Chute

Manual Start, Automatic Stop

Installation B
Pressure Plate Switch

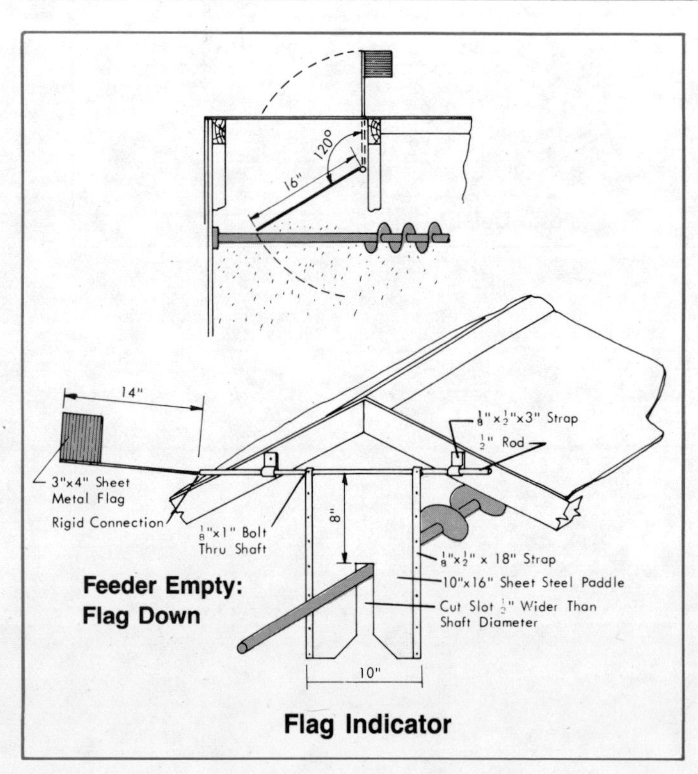

Flag Indicator

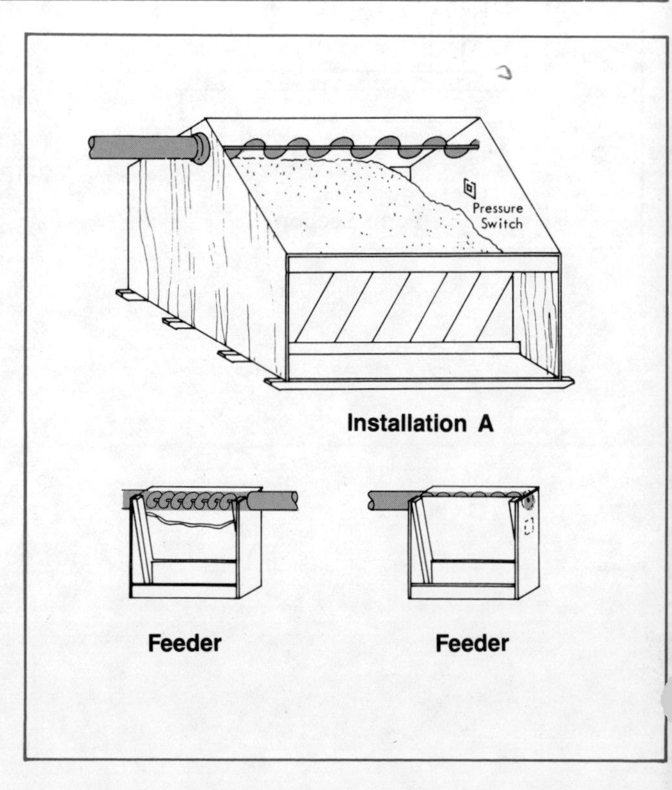

Installation A

Feeder

Feeder

Waterer

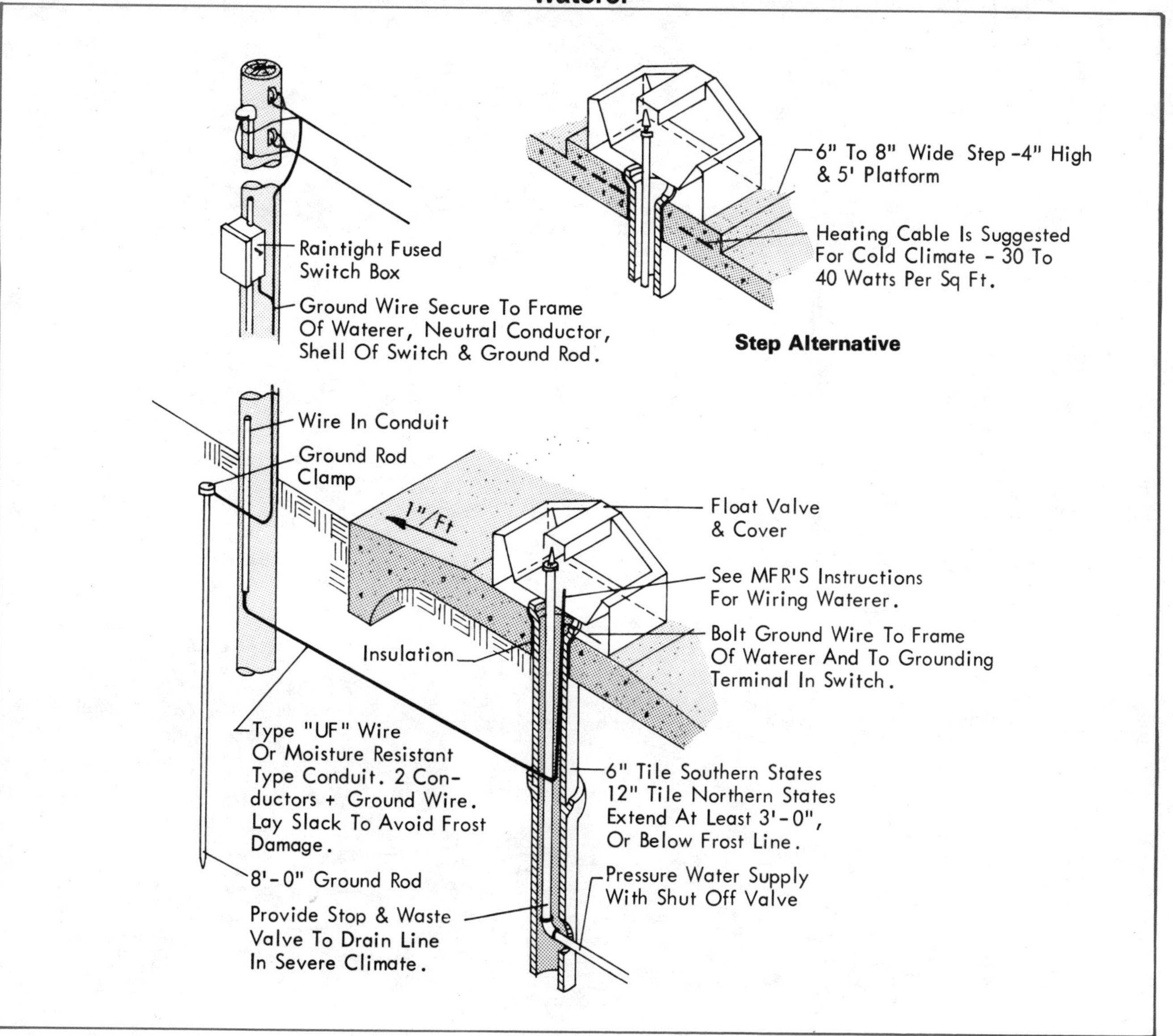

Barrel Waterer

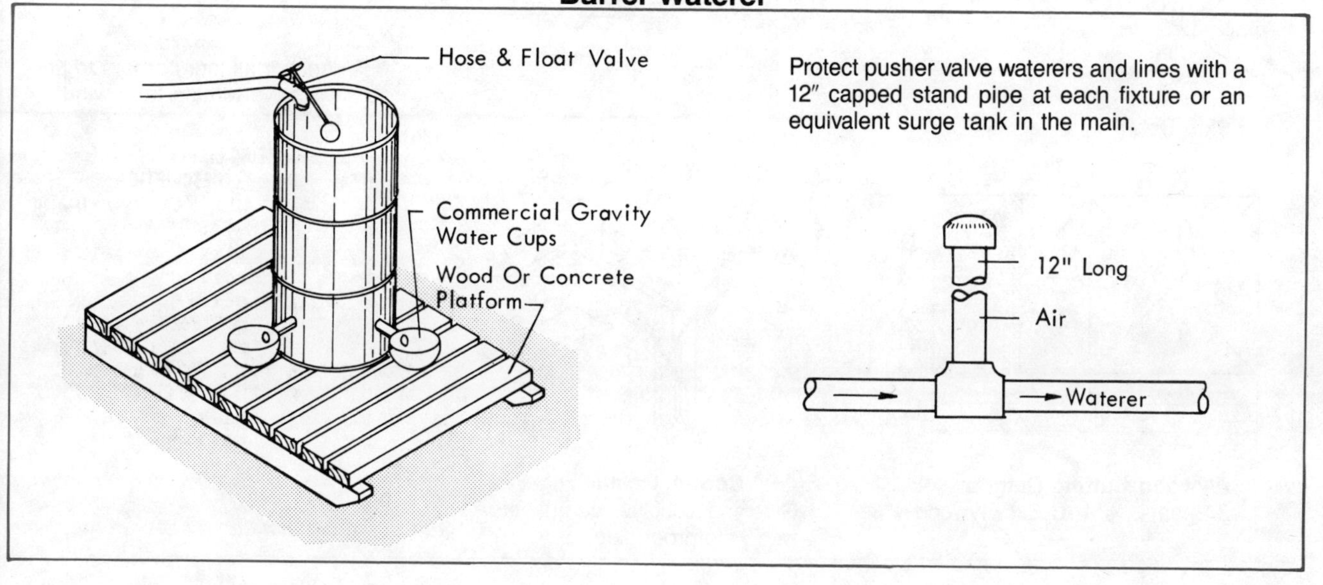

Sunshades

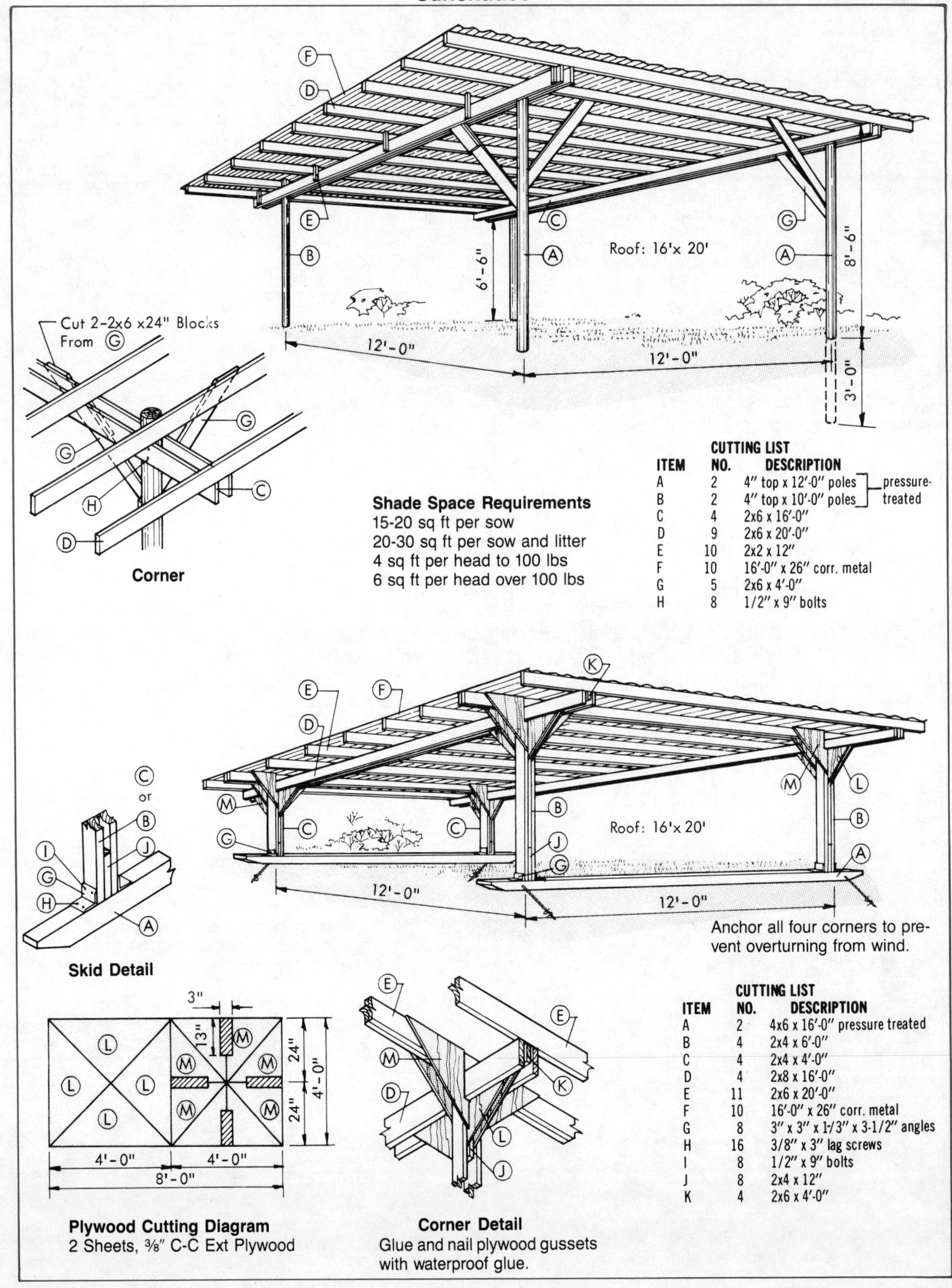

Shade Space Requirements
15-20 sq ft per sow
20-30 sq ft per sow and litter
4 sq ft per head to 100 lbs
6 sq ft per head over 100 lbs

CUTTING LIST

ITEM	NO.	DESCRIPTION
A	2	4" top x 12'-0" poles (pressure-treated)
B	2	4" top x 10'-0" poles (pressure-treated)
C	4	2x6 x 16'-0"
D	9	2x6 x 20'-0"
E	10	2x2 x 12"
F	10	16'-0" x 26" corr. metal
G	5	2x6 x 4'-0"
H	8	1/2" x 9" bolts

CUTTING LIST

ITEM	NO.	DESCRIPTION
A	2	4x6 x 16'-0" pressure treated
B	4	2x4 x 6'-0"
C	4	2x4 x 4'-0"
D	4	2x8 x 16'-0"
E	11	2x6 x 20'-0"
F	10	16'-0" x 26" corr. metal
G	8	3" x 3" x 1-1/3" x 3-1/2" angles
H	16	3/8" x 3" lag screws
I	8	1/2" x 9" bolts
J	8	2x4 x 12"
K	4	2x6 x 4'-0"

Plywood Cutting Diagram
2 Sheets, 3/8" C-C Ext Plywood

Corner Detail
Glue and nail plywood gussets with waterproof glue.

Sunshades

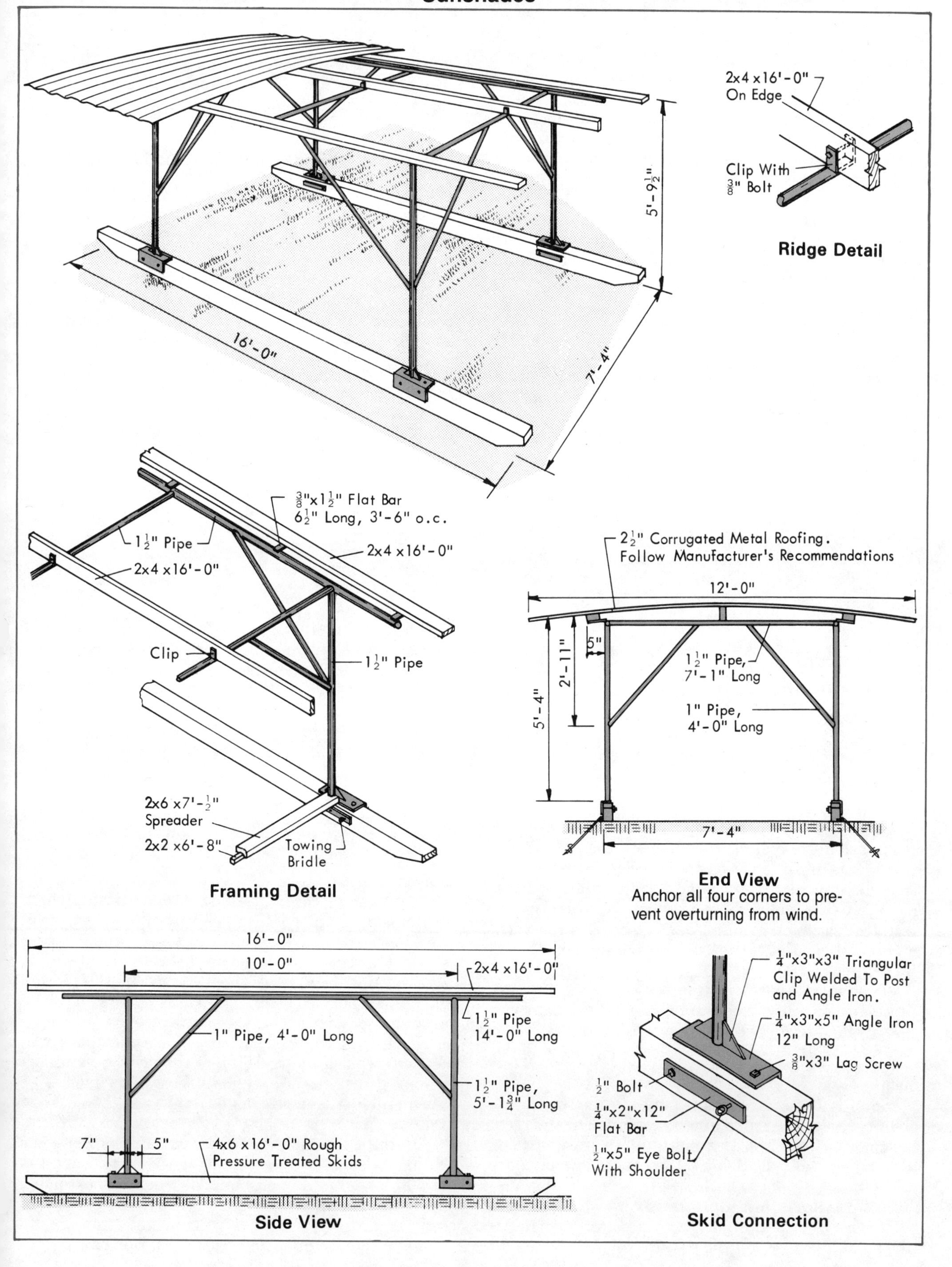

5'-9½"
16'-0"
7'-4"
2x4 x16'-0"
On Edge
Clip With
⅜" Bolt
Ridge Detail
⅜"x1½" Flat Bar
6½" Long, 3'-6" o.c.
1½" Pipe
2x4 x16'-0"
2x4 x16'-0"
Clip
1½" Pipe
2x6 x7'-½"
Spreader
2x2 x6'-8"
Towing
Bridle
Framing Detail
2½" Corrugated Metal Roofing.
Follow Manufacturer's Recommendations
12'-0"
5"
2'-11"
5'-4"
1½" Pipe,
7'-1" Long
1" Pipe,
4'-0" Long
7'-4"
End View
Anchor all four corners to prevent overturning from wind.
16'-0"
10'-0"
2x4 x16'-0"
1½" Pipe
14'-0" Long
1" Pipe, 4'-0" Long
1½" Pipe,
5'-1¾" Long
7"
5"
4x6 x16'-0" Rough
Pressure Treated Skids
Side View
¼"x3"x3" Triangular
Clip Welded To Post
and Angle Iron.
¼"x3"x5" Angle Iron
12" Long
⅜"x3" Lag Screw
½" Bolt
¼"x2"x12"
Flat Bar
½"x5" Eye Bolt
With Shoulder
Skid Connection

BUILDING AND EQUIPMENT MATERIALS

Wood Preservatives

Use commercially preservative-treated lumber where it will be exposed to soil, insects, manure, weather, or humid conditions. Brushing or dipping with oil- or water-soluble solutions is seldom worthwhile for these conditions. Select treatment by the expected exposure:

- Main framing exposed to soil, manure, or termite attack, and other lumber expensive to replace if the wood fails: creosote, 10 pcf; penta, 0.50 pcf; ACC salts, 0.50 pcf; ACA or CCA (Type A or B) salts, 0.40 pcf.
- Sills on foundations, framing less exposed or less expensive to replace, fence posts: creosote, 8 pcf; penta, 0.40 pcf; ACC salts, 0.25 pcf; ACA or CCA (Type A or B) salts, 0.25 pcf. (ACA = Ammoniacal Copper Arsenate; ACC = Acid Copper Chromate; and CCA = Chromated Copper Arsenate.)

Plywood

Use exterior-type plywood outdoors, inside animal shelters, or where alternate wetting and drying may occur. It has waterproof glue lines and all plies are at least grade C. Sheathing has waterproof glue lines and some grade D plies. Do not use sheathing outdoors unless it is covered with roofing or siding.

Paint and Finishes

Exterior Wood Finishes

Use stains on exterior plywood, low grade lumber, and rough or severely weathered wood. Paints frequently crack and peel on low-grade lumber in moist livestock housing.

Paint

Apply the appropriate primer. On redwood and cedar use an alkyd-oil base primer resistant to extractive staining. On woods not susceptible to extractive staining, like pine, use a high quality latex primer. Use alkyd-oil base primers with high moisture resistance on medium density hardboard. High quality latex paint can be both primer and topcoat on MDO plywood.

Topcoats: Apply **two** topcoats over the primer, especially on fully exposed areas. One topcoat may last 3 years—two topcoats of good quality paint can last as long as 10 years.

Stains

Semi-transparent stain (penetrating) is a latex or oil base stain. A small amount of pigment changes the color of wood but does not hide its grain or texture. It is usually available in natural and wood tone colors.

Solid color stain (opaque) is latex or oil base with more color and higher hiding power. It is more like paint and obscures the natural color of the wood. It comes in many colors, from natural wood tones through pastel to deep tone colors. The latex is easy to clean, is slower wearing, and has better color retention.

Interior Finishes for Wood and Wood Products

Penetrating nonfilm-forming

Penetrating wood sealers protect against dirt and stain without coloring the wood. They do not crack, peel, or scale, even if moisture gets into the wood.

Semi-transparent penetrating latex stains provide some protection from dirt and stain and add color. Semi-transparent oil base stains and penta are satisfactory, but do not use where contacted by livestock.

Surface film forming

Flat oil, latex, enamel, and paints form surface films. Some latex enamel manufacturers do not recommend their use on wood—follow manufacturer recommendations. The oil base or alkyd enamels with recommended sealer, primer, and undercoat are probably the most durable and cleanable film-forming finishes.

Painting Metals

Iron base metals

Painting metals successfully involves surface preparation, priming, and topcoating—preferably with compatable materials from the same manufacturer.

Tractor and implement paint, enamel or lacquer, exterior enamel, and exterior house paint can all be used. Avoid paints and primers that contain lead on surfaces exposed to animals. Water base or latex paints are less resistant to moisture than oil base paints and depend on the primer to prevent rust. Select a high quality primer intended for metal products that is compatible with the topcoats.

Galvanized metal

Galvanizing covers the base steel with zinc, which protects it from rusting. However, it eventually weathers away, the steel is exposed, and rusting starts. Painting galvanized metal helps maintain the zinc coating and extends the life of the metal. Paint a galvanized surface before any rust appears, or at the first sign of rust for best results.

If a new galvanized surface is to be painted immediately, pretreat it according to the primer manufacturer's directions. Apply metallic zinc primer after the metal has been pretreated or has weathered 3 to 4 years.

Metallic zinc dust paint weighs 20 lb or more per gallon. Although relatively expensive, it covers 400 to 500 ft^2/gal and is economical when the extended life of the metal is considered. It is usually battleship gray and can be used as a primer and a topcoat. To change color, apply a topcoat of exterior paint.

Aluminum

Wash bare aluminum with aluminum wool and a solvent wash. Avoid steel wool. Rinse thoroughly. Prime with zinc yellow primer. Topcoat the primer with two coats of paint or machinery enamel.

For repainting, clean and rinse as for bare aluminum. Lightly sand glossy areas with 100 to 150 grit open coat sandpaper. Rinse or wipe off dust.

Nails

Use galvanized nails where corrosion and staining may occur. Use annular- and spiral-grooved nails for structural joints. Cement coated nails increase joint strength for a short time, but strength drops to that of a plain nail joint in a few months.

Table 34. Size and strength of common nails.

Size	Length in.	Diam. in.	Approx. no/lb	Approx. strength lb Pull[1]	Lateral[2]
Common Nails					
2d	1	0.072	847	Douglas fir, larch or southern pine	
3d	1¼	0.080	543		
4d	1½	0.099	294		
5d	1¾	0.099	254		
6d	2	0.113	167	29	63
7d	2¼	0.113	150		
8d	2½	0.131	101	34	78
9d	2¾	0.131	92		
10d	3	0.148	69	38	94
12d	3¼	0.148	63	38	94
16d	3½	0.162	49	42	107
20d	4	0.192	31	49	139
30d	4½	0.207	24	53	154
40d	5	0.225	18	58	176
50d	5½	0.244	14	63	202
60d	6	0.263	11	68	223
Spikes					
10d	3	0.192	32	49	139
12d	3¼	0.192	31	49	139
16d	3½	0.207	24	53	155
20d	4	0.225	19	58	176
30d	4½	0.244	14	63	202
40d	5	0.263	12	68	223
50d	5½	0.283	10	73	248
60d	6	0.283	9	73	248
5⁄16	7	0.312	6	80	289
⅜	8-12	0.375	5-3	96	380
Hardened threaded nails					
6d	2	0.120	190	80	69
8d	2½	0.120	117	90	82
10d	3	0.135	78	100	94
12d	3¼	0.135	73	100	94
16d	3½	0.148	57	110	107
20d	4	0.177	36	135	139
30d	4½	0.177	31	135	139
40d	5	0.177	27	135	139
50d	5½	0.177	23	135	139
60d	6	0.177	18	135	139

[1]Per inch penetration of point.
[2]For penetration of 11 diameters.

Table 35. Nail selection.

Use galvanized hardened threaded nails for most wood outdoor projects. Use aluminum nails with aluminum sheets. The nailing strength of plywood is about the same as solid wood, but the greater resistance to splitting when nailed near edges is a definite advantage.

	Nail to use
1″ wood	8d
2″ wood	16d to 20d
3″ wood	40d to 60d
Concrete forms	Common or double headed nails
Toenailing studs	10d
Sheathing: roof, wall, and floor	8d
Roofing	
Aluminum	1¾″ to 2½″ aluminum nail with rubber washers.
Asphalt shingles	Large head roofing nail
Wood shingles	3d to 4d
Nailing steel sheet metal (roofing and siding)	Self-tapping screws, helical drive screws with lead washers.
Nailing to concrete	Concrete or cement nails or helical drive nails or drive bolts.
Plywood—Cabinet work	
¾″ plywood	8d finishing nails
⅝″ plywood	6d or 8d finishing nails
½″ plywood	4d or 6d
⅜″ plywood	3d or 4d
¼″ plywood	¾″ or 1″ brads; 3d nails
Plywood-Structural work	
Combined subfloor and underlayment: ¾″ or less	6d deformed shank
⅞″ or larger	8d deformed shank
Subflooring: ⅞″ or less	8d common
1″ or larger	10d common
Underlayment	3d ring shank
Roof sheathing: ½″ or less	6d common, ring shank
½″ or less	or spiral thread
⅝″ to 1″	8d common, ring shank or spiral thread

Concrete

Concrete is a mixture of portland cement, water, and aggregates. It is durable only if properly proportioned, mixed, and placed.

The cement and water form a paste that hardens and glues the aggregates together. The quality of concrete is directly related to the binding qualities of this cement paste.

Construction

Remove all sod and organic matter from the site. The subgrade must have uniform soil compaction and moisture content, and be easily drained. The top 6″ of subgrade should be sand, gravel, or crushed stone where it will be water soaked much of the time.

Use 6x6, 10x10 wire mesh in pavements poured over a nonuniform base. Place the wire about 1″ from the bottom of the slab. Place the wire near the top of slabs that will support light loads but are subject to frost heaving.

Thoroughly dampen the subgrade or place 6 mil plastic over the subgrade. Place concrete as near its final position as possible to reduce labor and segregation of the mix. Spade and vibrate along the forms to eliminate voids or honeycombs.

Table 36. Concrete mixes.
Make a trial batch to check for slump and workability. Wet sand occupies more volume than dry sand, so yield is not quite the same as in the example below.

		Gallons of water for each sack of cement, using:[1]			Suggested mixture for 1-sack trial batches[2]			Ready-Mix sacks cement per yard[6]
	Max. size aggregate	Damp[3] sand	Wet[4] (average) sand	Very[5] wet sand	Cement, sacks ft³	Aggregates Fine ft³	Aggregates Coarse ft³	
5-Gallon mix; use for concrete subjected to severe wear, weather, or weak acid and alkali solutions.	3/4"	4½	4	3½	1	2	2¼	7¾
6-Gallon mix; use for floors (home, barn), driveways, walks, septic tanks, storage tanks, structural concrete.	1"	5½	5	4½	1	2¼	3	6¼
	1½"	5½	5	4½	1	2½	3½	6
7-Gallon mix; use for foundation walls, footings, mass concrete etc.	1½"	6¼	5½	4¾	1	3	4	5

[1]Increasing the proportion of water to cement reduces the strength and durability of concrete. Adjust the proportions of the batches without changing the water-cement ratio. Reduce gravel to improve smoothness; reduce both sand and gravel to reduce stiffness.
[2]Proportions vary slightly depending on gradation of aggregates.
[3]Damp sand falls apart after being squeezed in the palm of the hand.
[4]Wet sand balls in the hand when squeezed, but leaves no moisture on the palm.
[5]Very wet sand has been recently rained on or pumped.
[6]Medium consistency (3" slump). Order air-entrained concrete for outdoor use.

Table 37. Air content for air-entrained concrete.

Max. aggregate size	Amount of air, %
1½", 2", or 2½"	4 to 6
¾" or 1"	5 to 7
⅜" or ½"	6½ to 8½

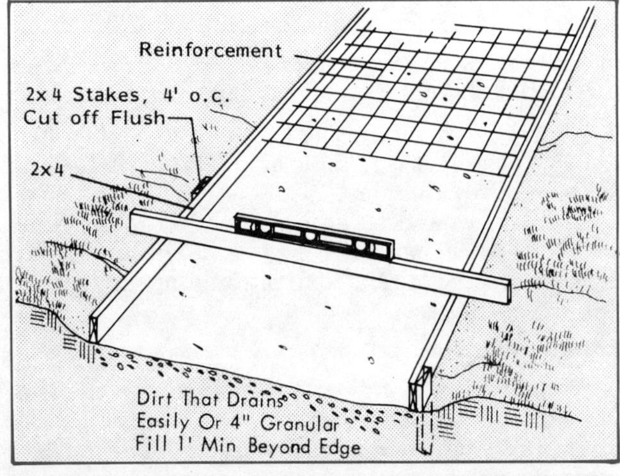

Fig 86. Preparing subgrades and laying reinforcing if needed.

Place concrete as soon as possible after it is mixed. Then strike it off to the proper grade; float to smooth and level the surface before any bleeding occurs. Bleeding is excess water in the concrete rising to the surface.

After all bleeding water has evaporated and the concrete has started to stiffen, round the edges to prevent chipping. Shrinkage cracks are unavoidable so cut control joints one-fourth of the slab thickness deep to prevent random cracks. Loads are transferred across the joints by the aggregates in the broken concrete surfaces below the cut. Divide the slab into rectangles:

- 4" thick-8'x12'.
- 5" thick-10'x15'.
- 6" thick-12'x18'.

Cut control joints in fresh concrete with a pointed trowel or straight hoe, or saw them after the concrete has cured enough for smooth cuts but before the random cracks form.

Float the slab again to embed large aggregates beneath the surface, remove slight imperfections and tool marks, and prepare the surface for other finishing.

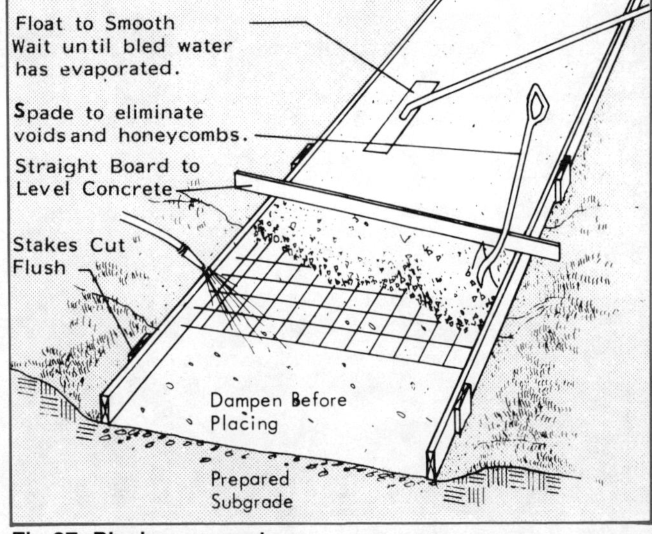

Fig 87. Placing concrete.

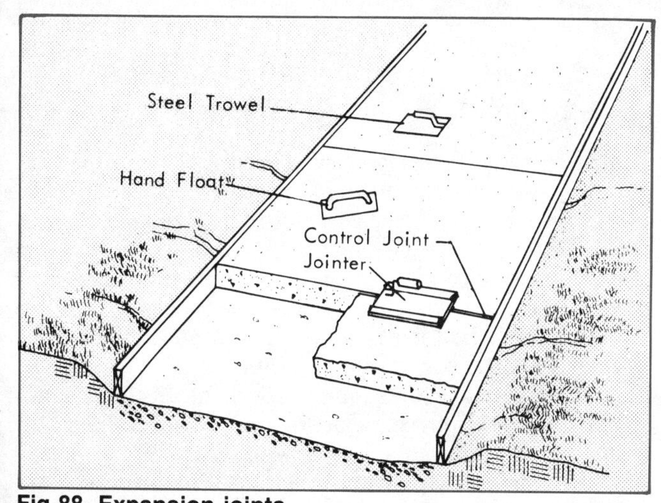

Fig 88. Expansion joints.
Expansion joints prevent buckling of long slabs in hot weather.

Steel trowel immediately after floating for a smooth surface. For a rough surface, omit steel troweling and float or broom the surface to the desired texture.

Isolation and expansion joints permit the slab to move with the earth and temperature changes. Place ¾″ wide isolation joints along existing improvements such as buildings, concrete water tanks, or paved drives. Install expansion joints in long walks and drives to prevent buckling of the slab during hot weather.

Expansion joints are the same as isolation joints and are installed in long walks and drives to prevent buckling of the slab during hot weather.

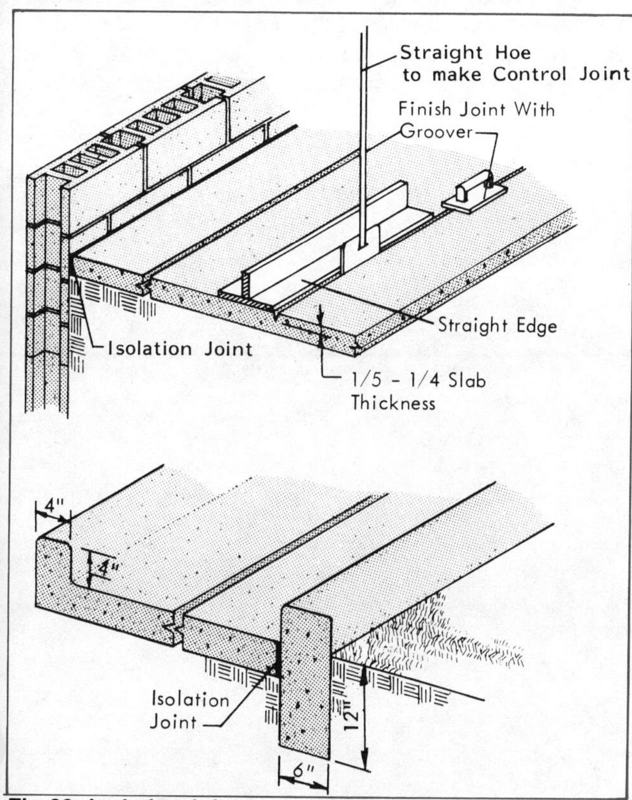

Fig 89. Isolation joints.
Isolation joints allow the slab to move with the earth.

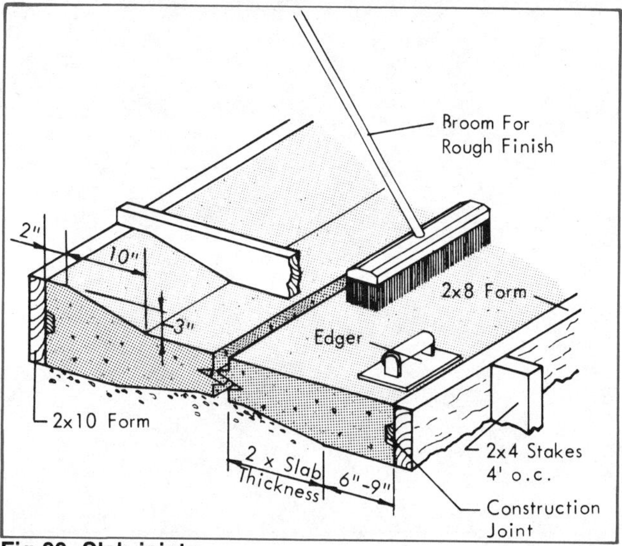

Fig 90. Slab joints.
Edges of slabs are thickened to reduce cracking and to allow for some erosion of adjacent soil. Construction joints key adjacent slab pours to each other.

Curing

Concrete does not dry—the paste sets by a chemical reaction between cement and water. Keep the surface of the concrete damp with curing compound, plastic cover, or wetted straw for at least 5 days. Curing continues for months. Remove forms after about 5 days for slabs, 10 days for walls, and 28 days for structural elements.

Floor Thickness

- 4″: Feeding aprons and floors with minimum vehicle traffic. Building floors.
- 5″: Paved feedlots, building driveways.
- 6″: Heavy traffic drives (grain trucks and wagons).

Slip Resistant Concrete Floors

Steel troweling brings fine aggregate and cement to the top, forming a smooth surface. Wood float and broom finished surfaces become smooth in time due to tractor scraping and constant animal traffic. Select the degree of roughness based on the type of animal.

To roughen new floors, score the surface with a homemade tool or add aluminum oxide grit after floating. Be certain the surface is hard enough not to let the concrete flow into the grooves or cover the grit.

Deep grooves make cleaning and disinfecting more difficult and may cause foot and leg problems with smaller animals. Make grooves diagonal to the direction of animal traffic. See Fig 91 and 92 for concrete grooving tools.

Paint existing slick concrete with chlorinated rubber paint **or** with an exterior latex—for additional traction sprinkle with sawdust or course ground cornmeal while paint is wet. Repaint annually.

For a gritty surface, apply aluminum oxide grit (as in sandpaper) at ¼ to ½ lb/ft^2 before the concrete sets. Coarse grit (4 to 6 meshes/in.) is recommended.

Refer to *Slip Resistant Concrete Floors,* AED-19 by Midwest Plan Service, for more information.

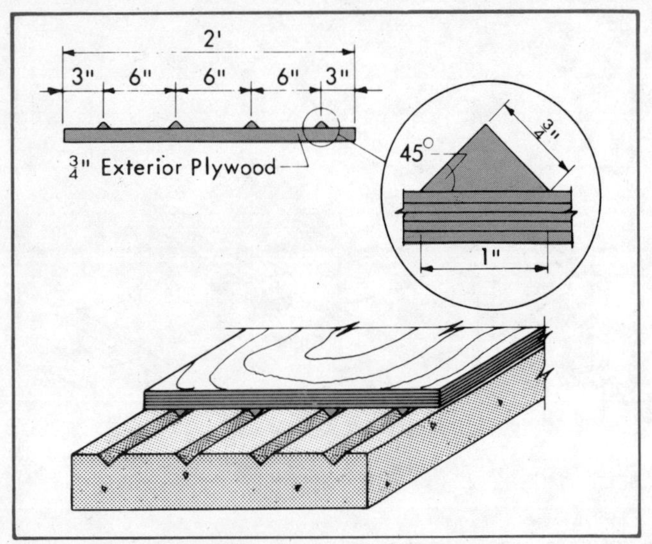

Fig 91. Wood grooves.

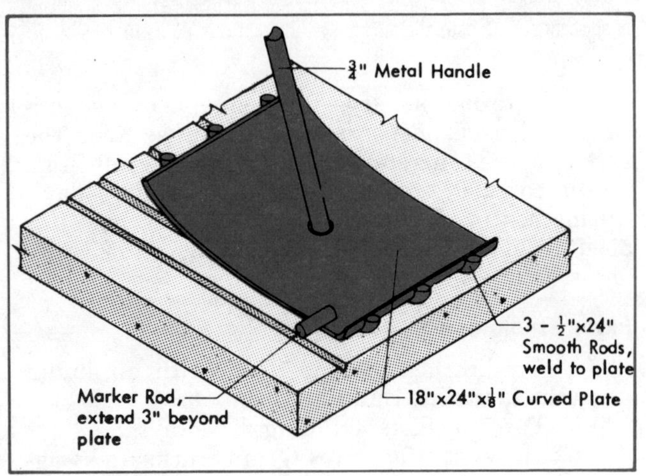

Fig 92. Steel grooves.

Glue

A glued joint is usually stronger than the wood it holds together. Use dry, smooth wood free of dirt, oil, and other coatings. Most purchased lumber has been planed and is sufficiently smooth. Clean off dirt, paint, and other coatings from the wood. Do not use wood with oil or grease at joint locations.

Generally, preservative-treated wood must be planed prior to gluing for maximum holding power. Wood treated with oil base preservatives tends to bleed; buy wood that has been steamed or otherwise cleaned until bleeding has stopped.

Two glues are recommended for buildings and livestock equipment: Casein and Resorcinol Resin. Follow manufacturers recommendations.

- Resorcinol Resin can be used for both wet and dry conditions. Buy a glue that meets Military Specification Mil-A-46051. Apply glue at 70 F or above. Assemble but wait 5 to 10 minutes before applying pressure. Maintain pressure for 10 to 16 hours.
- Use Casein only for dry or occasionally wet conditions. Buy a glue that meets Federal Specification MMM-A-125 Type II. Apply glue at 40 F or above; 70 F is recommended. Apply pressure as soon as possible. Maintain pressure for two days at 40 F, 4 hours at 70 F, or 2 hours at 80 F.

Pressure

Pressure squeezes glue into a thin continuous film, forces air from the joint, brings glue and wood together, and holds them until the glue has set and cured. Use clamps, nails, screws, or other fasteners before wiping off excess glue. Use box, galvanized, or cement coated nails—one per 8 in^2 of joint. Do not remove the nails after the glue has cured.